WATER TREATMENT PROCESSES

Simple Options

New Directions in Civil Engineering
Series Editor: W.F. Chen, Purdue University

Response Spectrum Method in Seismic Analysis and Design of Structures
A.K. Gupta, North Carolina State University

Stability Design of Steel Frames
W.F. Chen, Purdue University and E.M. Lui, Syracuse University

Concrete Buildings: Analysis for Safe Construction
W.F. Chen, Purdue University and K.H. Mossallam, Ministry of Interior, Saudi Arabia

Stability and Ductility of Steel Structures Under Cyclic Loading
Y. Fukumoto, Osaka University and G.C. Lee, State University of New York at Buffalo

Unified Theory of Reinforced Concrete
T.T.C. Hsu, University of Houston

Advanced Analysis of Steel Frames: Theory, Software, and Applications
W.F. Chen, Purdue University and S. Toma, Hokkaigakuen University

Flexural-Torsional Buckling of Structures
N.S. Trahair, University of Sydney

Analysis and Software of Cylindrical Members
W.F. Chen, Purdue University and S. Toma, Hokkaigakuen University

Buckling of Thin-Walled Structures
J. Rhodes, University of Strathclyde

Fracture Processes in Concrete
J.G.M. van Mier, Delft University of Technology

Fracture Mechanics of Concrete
Z.P. Bazant, Northwestern University and J.Planas, Technical University, Madrid

Aseismic Testing of Building Structures
B.-L. Zhu, Tongji University, Shanghai

Artificial Intelligence and Expert Systems for Engineers
C.S. Krishnamoorthy, Indian Institute of Technology, and S. Rajeev, Indian Institute of Technology

Limit Analysis and Concrete Plasticity, Second Edition
M.P. Neilsen, Technical University of Denmark

Simulation-Based Reliability Assessment for Structural Engineers
P. Marek, San Jose State University; M. Gustar, Prage; and T. Anagnos, San Jose State University

Winter Concreting
B.A. Krylov, Design and Technology Institute (NIIZhB), Moscow

Introduction to Environmental Geotechnology
H.-Y. Fang, Lehigh University

WATER TREATMENT PROCESSES

Simple Options

S. Vigneswaran
C. Visvanathan

CRC Press

Boca Raton   New York   London   Tokyo

Learning Resources
Centre

Library of Congress Cataloging-in-Publication Data

Vigneswaran, Saravanamuthu, 1952-
  Waater treatment processes : simple options / Saravanamuthu Vigneswaran and C. Visvanathan
      p.   cm.
  Includes bibliographical references and index.
  ISBN 0-8493-8283-1
  1. Water—Purification.   2. Appropriate technology.   I. Visvanathan, C.   II. Title.
  [DNLM: 1. Hepatitis B virus. QW 710 G289h]
  TD430.V54 1995
  628.1′62--dc20
  DNLM/DLC
  for Library of Congress   94-47108
                                                                                              CIP

Preface

Water is essential for human life. Its quantity and quality are equally crucial. However, natural waters are in most cases not aesthetically nor hygienically fit to be consumed directly, thus calling for some means of treatment.

The United Nations-sponsored water decade program (1981–1990) enabled many developing countries to adopt appropriate water treatment facilities, thus providing adequate quantities of safe water to their ever-increasing populations. During the same period, the potable water treatment and distribution sector has seen notable technical developments. Parallel to this progress, environmental science and engineering activities have also flourished. Many national universities have developed comprehensive undergraduate and postgraduate programs in environmental science and engineering. Drinking water treatment and supply is one of the key fields of study in such programs.

From the environmental engineering point of view, drinking water supply activities can be broadly divided into two groups: water distribution and water treatment. The former basically deals with the hydraulics of water conveyance from source to the treatment plant and the distribution of purified water to consumers. This field is still largely taught as part of the traditional civil engineering program. The literature in this field is extensive, and a great variety of computer software is also available.

This book therefore considers the latter type. In this field, the available books are concerned mainly with the theoretical concepts and the design of conventional and/or advanced treatment technologies. Very little emphasis has been given so far to low-cost or simple treatment technologies. The main objective of this book, therefore, is to bridge the gap, by discussing both conventional and simple treatment options.

Most of the water treatment technologies practiced in both developed and developing countries look simple but attention must be given to their cost and appropriateness. Technologies such as membrane filtration are certainly very effective but at present the cost involved is prohibitive. Therefore, when selecting a treatment technology, the appropriateness and the economy of the technology must be an integral part of the selection process.

It has become a standard practice in the small community water supply schemes to look for a treatment option which already exists. They are easily tempted by the glamour of hi-tech alternatives without considering the operation and maintenance complications that follow.

Quite a number of low-cost, appropriate treatment alternatives have been tried and tested in small community water supplies. Countries like Brazil, China, and India have developed several treatment techniques (traditional[1] and non-conventional[2]) which are extensively used.

---

1 Traditional techniques: Techniques commonly used. Example, rectangular sedimentation tank, slow sand filter, baffle type flocculator.
2 Non-conventional techniques: Techniques not commonly used but developed to suit the local situation. Example, inline blenders, Alabama type flocculator, tube settlers.

Public health engineering professionals have not given much thought to these innovative low-cost methods which are either already in use or are being tried with considerable success, at least on a pilot scale. For example, treatment technologies such as gravel-bed flocculators are overlooked in favor of more complex conventional technologies such as mechanical flocculators. There are hundreds of such low-cost simple treatment technologies which can successfully be applied in the developing world. One of the aims of this book is to highlight some of these simple treatment technologies which are already being used to some extent or those which look quite promising to be adopted. Emphasis is given to their application status and efforts are made to outline their design criteria in a simplified manner. Not all the technologies have a theoretical design concept behind them because of the complexity in the kinetics involved. Therefore, wherever necessary, design guidelines to suit the small to medium size communities are given instead of or in addition to the theoretical design concept. This will be effected through assessing relative merits and applicability status and detailed process designs of different technologies. Meeting drinking water quality standards (physical, chemical, and biological) will be considered as prime criteria in selecting the treatment options. The aspect of process modifications and upgrading of conventional treatment facilities is also discussed in detail

This book is addressed to three groups of people: final year undergraduate students specializing in water treatment; graduate students on environmental programs; and engineers and mangers involved in design and selection of appropriate treatment methods. The emphasis placed on practical aspects, with numerous figures, tables, and worked-out examples, enable this book to be used as a supplementary text book.

The first chapter in this book introduces the quantity and quality requirements in water supplies. The various methods of water quantity projections as well as assessment of the target population are discussed in detail. The section on water quality covers physical, chemical, and microbiological aspects of water quality.

The second chapter which is on water sources and treatment technologies outlines both conventional and advanced water treatment processes. The advanced processes include flotation, adsorption, ion exchange and membrane processes. While these two chapters discuss the overall view, the subsequent six chapters extensively discuss six unit processes in drinking water treatment.

Rapid mixers are discussed in the third chapter considering the principle, design application status, and relative merits of both mechanical and hydraulic mixers. The fourth chapter on flocculation introduces various types of flocculation devices available. Baffled channel flocculator, gravel-bed flocculator, and Alabama-type flocculator are discussed in detail, with emphasis on design concepts, application status and relative merits. The discussion on sedimentation given in Chapter 5 covers the design details, application status and relative merits of conventional sedimentation tanks, tube settlers and sludge blanket clarifiers.

The sixth chapter is on filtration. The first part considers the design, operation, and maintenance of slow sand filters. This is followed by a discussion on simple filtration and treatment facilities, namely horizontal flow coarse media filtration, two-stage filter, modified shore filtration and the SWS filter system. The second part of this chapter discusses rapid filtration. The design, operation, and maintenance of conventional filters are considered. The modifications of the rapid filter in terms of filter media, filtration rate, and flow direction are also included. The final part of this chapter considers direct filtration. Again, the design, economics, advantages and disadvantages are discussed along with the application status.

Water treatment for specific impurity removal is considered in the seventh chapter. Here the removal of iron, manganese and fluoride are discussed. The available methods, their relative merits, and applicability are considered.

The last chapter is on disinfection. Here the first part discusses chlorination, especially using bleaching powder as the disinfectant. Alternative disinfectants like chloramine, chlorine dioxide, ozone and UV radiation are considered in the second part.

Finally special thanks are due to Mr. D. R. Induka B. Werellagama, Mr. Biswadeep Basu, Mr. Huu Hao Ngo, Mr. Ngo Lee Seung Hwam, and Mr. V. Jegatheesan for their kind assistance in reviewing the manuscript, examples, and drawing some of the figures.

S. Vigneswaran, University of Technology, Sydney
C. Visvanathan, Asian Institute of Technology, Bangkok

Contents

1: Water Quantity and Quality

CONTENTS

1.1 INTRODUCTION

The United Nations International Drinking Water and Sanitation Decade 1981–1990 set its goal to provide for safe drinking water and sanitation for all the world's people (more than two-thirds of whom live in rural areas). For the 1 billion people in developing countries who do not have access to clean water and the 1.7 billion who lack access to sanitation, these are the most important environmental problems of all. The economic costs of inadequate provision are also high. Many women in Africa spend more than two hours per day fetching water. In Jakarta an amount equivalent to 1% of the city's gross domestic product (GDP) is spent each year on boiling water, and in Bangkok, Mexico City, and Jakarta, excessive pumping of groundwater has led to subsidence, structural damage, and flooding. These problems demand simple, sustainable, low-cost technology solutions for both urban and rural water supply. Community involvement is essential for development if successful implementation and future maintenance are to be ensured (WDR, 1992). For rural water supply schemes, community involvement is essential on at least three counts:

- to ensure commitment to use of the scheme
- to mobilize village resources in terms of manpower, goods and services
- to ensure that sound arrangements can be instituted for long-term maintenance

The plight of two-thirds of the world's population insofar as it is exposed to the dangers of unsafe water supply was highlighted by the publicity that the United Nations International Drinking Water and Sanitation Decade generated. The effects of inadequate water and sanitation on health are shocking; these are major contributors to the 900 million cases of diarrheal diseases every year, that cause the death of more than 3 million children. Two million of these deaths could have been prevented if adequate sanitation and clean water were available. At any time, 200 million are suffering from schistosomiasis or bilharzia and 900 million from hookworm. Cholera, typhoid, and paratyphoid also continue to wreak havoc with human welfare. Providing access to sanitation and clean water would not eradicate all these diseases, but it would be the single most effective means of alleviating human distress (WDR, 1992).

Comparison of statistics indicates progress in rural water supply in terms of percentage of population supplied with water, but there is some regression in urban water supply mainly because of population drift from rural to urban areas. Domestic water use in developing countries will rise sixfold over the coming four decades. The bulk of demand will come from urban areas, where populations will triple (WDR, 1992). This increase will place severe strains on surface and groundwater supplies.

Therefore, the objectives of water supply schemes are the following:

- to provide safe water in adequate quantity
- to locate water supply facilities with easy access
- to make water available at a reasonable cost

The basic factors to be considered in water supply schemes are

- area and population to be served
- design period
- water demand both with regard to quality and quantity
- selection of water sources (e.g., groundwater, surface water)
- nature and location of transmission and distribution system
- economic aspects

Cost of the water supply scheme depends on:

- selection of water sources
- treatment method adopted
- nature and extent of water distribution

On average, households in developing countries pay only 35% of the cost of supplying water. The vast majority of urban residents want in-house supplies of water and are willing to pay the full cost. Yet many countries have assumed that people cannot afford to pay the full costs, and they have therefore used limited public funds to provide a poor service to a restricted number of people (WDR, 1992).

1.2 WATER QUANTITY

1.2.1 Water Requirement

Water consumption is commonly referred as the amount of water taken from distribution systems; however, little of it is actually consumed and most of it is discharged as wastewater. The water demand of a community depends on:

- climate
- standard of living
- type and extent of sewerage system used
- water pricing (complete meterage)
- availability of private supply
- method of distribution

Depending on the climate and work load, the human body needs about 3–10 liters of water per day for normal functioning (IRC, 1981). The different types of urban water usage for domestic purposes are given in Table 1.1(a) for India and other South Asian countries. This does not include water for gardening, livestock, etc. Typical domestic consumptions for Melbourne, Australia are given in Table 1.1(b) for comparison. An allowance of 10–30% of the total water demand should be considered for water losses and wastage.

Table 1.1(a)   Water Uses for Domestic Purposes in Southern Asia (Design Manual for Water Supply and Treatment, India, 1991)

Purpose	Quantity (Lcd)[a]
Drinking	5
Cooking	3
Sanitary purposes	18
Bathing	20
Washing utilities	15
Clothes washing	20
Total (excluding water loss and wastage)	81

[a]Lcd: Liters per capita per day.

Table 1.1(b)   Typical Domestic Consumptions for Melbourne, Australia (Adapted from Jolliffe, 1991)

Use	Quantity (Lcd)[a]	Percentage of total usage
Toilet, laundry	100	11
Washing/drinking	30	3
Shower/bath	30	7
Garden	700	78
Miscellaneous	10	1
Total	870	

[a]Lcd: Liters per capita per day.

While a minimum of 70 to 100 liters per capita per day (Lcd) may be considered adequate for the domestic needs of urban communities, the nondomestic needs of urban communities would significantly push up this figure (Table 1.1(b)). Per capita design consumption rates used by different agencies in New South Wales and Canberra in Australia are given in Table 1.2.

Table 1.2   Design Consumption Rates in Australia (Adapted from Jolliffe, 1991)

Agency	Average daily demand (Lcd)
Department of Public Works (PWD)	
Coastal and Tablelands	275
Western	340
Water Board	
Total system	550
Sydney	517
South Coast	1000
National Capital Development	
Corporation (NCDC) average	1700

Table 1.3 gives the typical domestic water usage data for different types of water supply systems in developing countries (IRC, 1981). It is clear from this table that the demand increases with the increase of service level.

The commercial and institutional water requirements in developing countries are given in Table 1.4(a). An allowance of about 20% for water losses and wastage is included. Water losses and wastage should be considered in water demand calculations as they are significant in quantity. Unaccounted-for water, much of it unused, constitutes as much as 58% of piped water supply in Manila and 40% in most Latin American countries (WDR, 1992). Table 1.4(b) shows the annual domestic, commercial, and institutional demands in New South Wales, Australia for comparison. The requirements of each case must be studied in detail before the water demand for the community is decided upon. The values reported in Tables 1.4(a) and 1.4(b) are only indicative ones.

Table 1.3   Typical Domestic Water Usage in Developing Countries (IRC, 1981)

Type of water supply	Typical water consumption (Lcd)[a]	Range (Lcd)[a]
Communal standpipe walking distance < 250 m	30	20–50
Yard connection (tap placed in house-yard)	40	20–80
House connection		
single tap	50	30–60
multiple tap	150	70–250

[a]Lcd: Liters per capita per day.

Table 1.4(a) Various Water Requirements
in Developing Countries (IRC, 1981)

Category	Typical water use
Schools	
Day schools	15–30 L/day per pupil
Boarding schools	90–140 L/day per pupil
Hospitals (with laundry facilities)	220–300 L/day per bed
Hostels	80–120 L/day per resident
Restaurants	65–90 L/day per seat
Cinema houses, concert halls	10–15 L/day per seat
Offices	25–40 L/day per person
Railway and bus stations	15–20 L/day per user
Livestock	
Cattle	25–35 L/day per head
Horses and mules	20–25 L/day per head
Sheep	15–25 L/day per head
Pigs	10–15 L/day per head
Poultry	
Chicken	15–25 L/day per 100

In assessing per capita demand, one needs to know domestic and small industry needs, institutional and major industry needs, fire fighting, requirements for livestock, and percentage of waste among all users. In assessing these needs, attention should be paid to the local needs, habits of the people and their living standards, and the industrial and commercial importance of the area. In the absence of specific data, as a preliminary estimation, a value of 1.5 L/s (or more) per 1000 people can be assumed for small community water supply schemes (IRC, 1981).

To allow for future population growth and a higher use of water per capita (or per household), a community water supply system must have sufficient surplus capacity. The design is typically based on the daily water demand estimated for the end of a design period or the demand computed on the basis of the population growth estimation.

Table 1.4(b) Annual Commercial and Institutional Demand
in New South Wales, Australia (Adapted from Jolliffe, 1991)

Use	Basic annual demands (m^3)
Commercial demands	
Hotel/motel	1 permanently occupied house + 50 m^3 per unit
Caravan park and camp sites	60 per site
Offices, shops	
Minor establishments	10,000 per ha
Major establishments	Individually assessed
Institutional demands	
School	10 per pupil
Hospitals	350 per bed
Nursing homes and similar	100 per bed
Public places	
Parks, gardens, golf courses, etc.	250 per ha per week of watering

Table 1.5 Design Period for Different Components of Water Supply Schemes (Design Manual for Water Supply and Treatment, 1991)

Item	Design period in years
1. Storage by dams	50
2. Infiltration works	30
3. Pump sets:	
(i) all prime movers except electric motors	30
(ii) electric motors and pumps	15
4. Small community water treatment units	15
5. Pipe connection to the several treatment units and other small appurtenances	30
6. Raw water and clear water conveying mains	30
7. Clear water reservoirs at the head works, balancing tanks, and service reservoirs (overhead or ground level)	15
8. Distribution system	30

In any water supply project, the design period is fixed to compute the design population and thus the design demand. Generally, different components in the system are designed for different periods, as given in Table 1.5.

1.2.2 Population Projection

The design population or population projection at the end of the design period is calculated by various methods as listed below.

- census data
- combination of census and zoning plan of councils
- population trends (Figures 1.1 and 1.2 present sample population trends for Sydney and selected cities in New South Wales)
- regression method

Assuming Geometrical Progression

$$P_n = P_0 \left(1 + \frac{r}{100}\right)^n$$

where P_n = design population, P_0 = initial population, r = growth rate in percentage (about 2–3% per year for Southeast Asia), n = design period.

Assuming Incremental Increase

$$P_n = P_0 + nX + n\frac{(n+1)}{2}Y$$

where X = average increase of population, Y = incremental increase of population.

Logistic Method

$$P_n = \frac{L}{\left(1 + me^{nM}\right)}$$

where L = upper limit on population, m and M = constants to be evaluated.

A worked-out example on the calculation of design population is given at the end of this chapter.

1.3 WATER QUALITY

The majority of water supplies require treatment to make them suitable for use in domestic and industrial applications. Although appearance, taste, and odor are useful indicators of the quality of drinking water, suitability in terms of public health is determined by microbiological, physical, chemical, and radiological characteristics. Of these, the most important is microbiological quality. Also a number of chemical contaminants (both inorganic and organic) are found in water. These cause health problems in the long run and, therefore, detailed analyses are warranted.

The drinking water thus should be

- free from pathogenic (disease causing) organisms
- clear (i.e., low turbidity, little color)
- not saline (salty)
- free from offensive taste or smell
- free from compounds that may have adverse effects on human health (harmful in the long term)

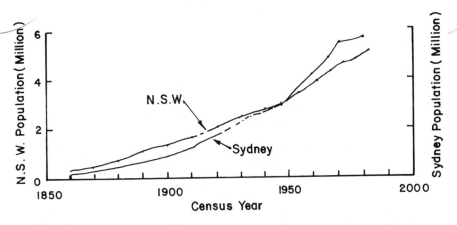

FIGURE 1.1 Population trends for Sydney and New South Wales. (Adapted from Jolliffe, 1991)

- free from chemicals that may cause corrosion of water supply system or stain clothes washed in it

To ensure safe drinking water, detailed water quality standards have been proposed by different countries and international organizations. These guidelines provide the following information for water authorities, health officials, and consumers:

- day-to-day operational values to ensure that the supplied water does not carry any significant risk to the consumer
- a basis for planning and designing water supply schemes
- assessment of long-term trends of the performance of the system

The guideline values are broadly classified into physical, chemical, microbiological, and radiological characteristics. These characteristics are discussed in this chapter under separate headings together with the World Health Organization (WHO), the United States, the European Economic Community (EEC), Canadian, and Australian guidelines.

A total of 42 items are designated by WHO in the evaluation of water quality so far. While U.S. national guidelines recommend the examination of 43 items, the U.S. Environmental Protection Agency (U.S. EPA) proposes 250, and the new Japanese standard, amended in 1993 proposes 85 (Chung, 1993).

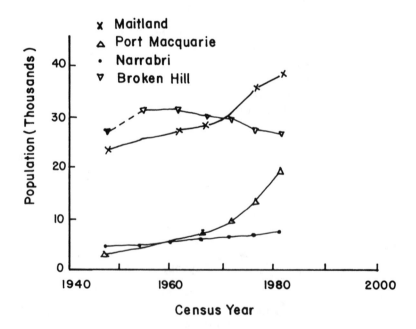

FIGURE 1.2 Population patterns for selected New South Wales cities. (Adapted from Jolliffe, 1991)

1.3.1 Physical and Chemical Aspects

Although in small community water supplies (especially in developing countries), the water quality problems are mainly due to bacteriological contamination, a significant number of very serious problems may occur as a result of chemical contamination of water resources. Over 60,000 potentially harmful chemicals are now being used by various segments of industry and agriculture (approximately 10,000 chemicals are listed in Korea, 30,000 in Japan, and 60,000 in the U.S. (Chung, 1993)).

In order to maintain the water quality, important physicochemical parameters need to be measured. The important parameters are discussed below, together with the respective maximum allowable concentration limits.

Organoleptic Parameters

High turbidity and/or color imparts an aesthetically displeasing appearance to water. The turbidity in surface waters results from the presence of colloidal material (such as clay and silt), plankton, and microorganisms. Apart from displeasing appearance, turbidity provides adsorption sites for chemicals that may be harmful or cause undesirable taste and odor. It also provides adsorption sites for biological organisms and interferes with disinfection.

Color in drinking water is due to natural organics such as humic substances or dissolved inorganics such as iron and manganese. Highly colored industrial waste also imparts color to the water. Apart from the unaesthetic appearance caused by color to water, the organic causing color, when disinfected with chlorine, will produce chlorinated organics that are carcinogenic.

Odor problems in water are mainly due to the presence of organic substances. Water supplied to consumers should be free of objectionable taste and odor. The maximum allowable concentrations of organoleptic parameters are summarized in Table 1.6.

Table 1.6 Standards for Physical Quality (Adapted from Sayre (1988), NHMRC-AWRC (1987), and WHO (1993))

Characteristic	U.S.	Canadian	EEC	WHO	Australian guideline
Turbidity (NTU)	1–5	5[a]	0–4	<5	5
Color (TCU)	15	15	20 mg Pt-Co/L	15	15
Odor (TON)	3	—	0–2 dilution numbers at 12°C	NS	no objectionable odor
pH	6.5–8.5	6.5–8.5	6.5–8.5[b]	6.5–8.0	6.5–8.5
Taste	—	—	2 dilution numbers at 12°C	—	—

Note: — = Data not available; NS = No Standard.

[a]Decker and Long (1992) give the Canadian Standard for turbidity as 5 NTU aesthetic objective and 1 NTU as the maximum acceptable value. [b]pH value 6.2–8.5 according to EC directive 80/778 (AQUA, 1992).

Table 1.7   Standards for Chemical Quality (inorganic, health-related) (Adapted from Sayre, (1988); and NHMRC-AWRC, (1987))

Characteristic	U.S. (maximum level, mg/L)	Canadian (maximum level, mg/L)	EEC (maximum admissible concentration, mg/L)	WHO guideline (mg/L)	Australian guideline (mg/L)
Arsenic	0.05	0.025[a]	0.05	0.01	0.05
Barium	1.0	1.0	0.1	0.7	NS
Cadmium	0.01	0.005	0.005	0.003	0.005
Chromium	0.05	0.05	0.05	0.05	0.05
Fluoride	4.0[b]	1.5	1.5	1.5	0.5–1.7
Lead	0.05	0.01[a]	0.05	0.01	0.05
Mercury	0.002	0.001	0.001	0.001	0.001
Nitrate (as N)	10	10	50 as NO_3^-	50 as NO_3^- 3 as NO_2^-	10
Selenium	0.01	0.01	0.01	0.01	0.01
Silver	0.05	0.05	0.01	NS	NS

Note: NS = no standard.

[a]0.025 mg As/L and 0.01 mg Pb/L (Decker and Long, 1992). [b]Values of 4 and 2 mg/L given (Pontius, 1992).

Tables 1.7 and 1.8 summarize various standards for inorganic substances (both health-related and nonhealth-related). These chemicals enter the aquatic environment through geological weathering, soil leaching, mining, agriculture, and industrial discharges. Arsenic at concentrations above 0.05 mg/L leads to adverse health effects. Fluorides at levels above 1.5 mg/L have been reported to cause mottling of teeth. Elevated concentrations of nitrate (>10 mg/L as N) cause methemoglobinemia in infants. High levels of nitrates are found in groundwaters. Cadmium, chromium, lead and mercury even at low concentrations can be toxic. They tend to accumulate in the food chain and lead to biomagnification.

Table 1.8   Standards for Chemical Quality (inorganic, not directly health-related) (Adapted from Sayre (1988) and NHMRC-AWRC (1987))

Characteristic	U.S. (maximum level, mg/L)	Canadian guideline (mg/L)	EEC guideline (mg/L)	WHO guideline (mg/L)	Australian guideline (mg/L)
Aluminum	—	0.2	—	0.2	0.2
Chloride	250	400	250	25[a]	250
Copper	1	1	1	0.1	2.0
Hardness (as $CaCO_3$)	—	500	—	—	500
Total dissolved solids	500	—	500	—	1000
pH	6.5–8.5	—	6.5–8.5	6.5–8.5	8.5
Fe	0.3	0.3	0.3	0.3	0.3
Mn	0.05	0.02	0.05	0.05	0.1
Cyanide	—	—	—	—	0.05

Note: — = not specified or not available.

[a]Guideline value; maximum admissible concentration value is not reported.

Table 1.9 Standards for Organic Quality (health-related) (Adapted from Sayre (1988) and NHMRC-AWRC (1987))

Characteristic	U.S.	Canada	EEC	WHO	Australia
Trihalomethane precursors (mg/L)	0.1	0.35	0.001	0.03 (CHCl$_3$ only)	0.2
Pesticides (mg/L)	NS	0.1	0.005[a]	[b]	NS
2,4 D (mg/L)	0.1	0.1	NS	0.1	0.1
Endrin (mg/L)	0.0002	0.0002	NS	NS	NS
Lindane (mg/L)	0.0004	0.004	NS	0.002	0.1
Methoxychlor (mg/L)	0.1	0.1	NS	0.02	NS
Polycyclic aromatic hydrocarbons (μg/L)	—	—	0.2	—	0.01[c]

Note: NS = not specified; — = not available.

[a]0.001 mg/L for each pesticide and 0.005 mg/L for the total amount of pesticides present in drinking water.

[b]Guidelines are given in Table A.2.2.C of WHO, 1993.

[c]Benzo-a-pyrene (an indicator of pollution by toxic polynuclear aromatic hydrocarbons).

Aluminum compounds are used in water treatment plants as coagulants. Water containing 0.2 mg/L of aluminum (or more) is not suitable for use by kidney dialysis patients. Both iron and manganese at high concentrations cause stains in laundry and plumbing fixtures and impart undesirable tastes to beverages. Presence of iron in water may result in undesirable growth of iron bacteria in distribution systems and can block pipes.

Total dissolved solids (TDS) in raw water are mainly due to sodium bicarbonate, chloride, calcium, and magnesium bicarbonates and sulfates. High TDS causes unacceptable taste in water. Taste thresholds vary widely depending on the type of dissolved solids present.

Table 1.9 presents different standards for selected organic substances. The details can be found elsewhere (Sayre, 1988; Carney, 1991; WHO, 1993; NHMRC-AWRC, 1987). In the drinking water quality standards for organic substances established before the 1980s, emphasis was given only to pesticides and polycyclic aromatic hydrocarbons (PAH). Since then, knowledge of contamination of organic substances in water has increased. Out of more than 600 organic contaminants identified in drinking water, several were found to be carcinogenic and a number of them have been shown to be mutagenic (WHO, 1984). However, these organics represent only a small fraction of the total organic matter present in drinking water. Table 1.10 presents selected items that are classified into the U.S. EPA carcinogenic assessment categories.

Table 1.11 presents a list of groups of organic compounds of potential health significance that can be found in water sources and in treatment and distribution systems. These compounds can be broadly classified into pesticides; organic from industrial sources like chlorinated alkanes, ethenes, and benzenes; aromatic hydrocarbons; disinfectants and disinfectant by-products like chlorophenol, trihalomethanes, etc.

Most of the pesticides degrade rapidly in the environment. However, the toxicity of the degradated products has not been considered in guidelines in most cases due to inadequate data. Some of the pesticides are very toxic, while others are not.

Table 1.10 Classified Categories of Carcinogens (Chung, 1993)

Category	A	B1	B2	C	D
Volatile organic compounds	Benzene vinyl chloride		Trichloroethylene Pentachlorophenol Chloroform Bromodichloro- methane bromoform Carbon tetrachloride 2,4,6-Trichlorophenol Acrylamide 1,2-Dichloroethane	1-Dichloroethane 1,1,1,2-Tetrachloro- ethylene 1,1-Dichloro- ethylene Dibromochloro- methane P-Dichlorobenzene	Toluene Xylene Mono-chlorobenzene Ethylbenzene Hexachloro- cyclopentadiene 1,1,1-Trichloroethane 1,2-Dichlorobenzene
Pesticides			Heptachlor epoxide Hexachlorobenzene Toxaphene Aldrin Malathion	Endrin Methoxychlor Ethoxychlor	
Heavy metals	Arsenic (As) (inorganic) Chromium (Cr^{+6}) (inorganic)	Cadmium (Cd)	Lead and compounds (Pb) (inorganic)		Cyanide (CN) Silver (Ag) Mercury (inorganic) Manganese (Mn) Zinc (Zn) Selenium (Se) Copper (Cu)

Note: A. Human carcinogen, based on sufficient evidence from epidemiological studies. B. Probable human carcinogen, based on at least limited evidence of carcinogenicity to humans (B1), or usually a combination of sufficient evidence in animals and inadequate data in humans (B2). C. Possible human carcinogen, based on limited or equivocal evidence of carcinogenicity in animals in the absence of human data. D. Not classifiable, based on inadequate evidence of carcinogenicity from animal data.

(Guidelines range from 0.03–100 mg/L.) The organic compounds from industrial sources are mainly from solvents and the petroleum industry. Some polynuclear aromatic hydrocarbons have been seen to be carcinogenic and, accordingly, the WHO recommends against the use of coal tar in pipe linings and water storage tanks (WHO, 1993).

Chlorination of drinking water containing natural organic substances such as humic and fulvic acids produces a series of by-products such as trihalomethanes

Table 1.11 Groups of Organic Compounds of Potential Health Significance That Could Be Present in Water Sources and Treatment and Distribution Systems (Adapted from WHO, 1993)

Chlorinated alkane	Chloramines
Chlorinated ethenes	Chlorophenols
Aromatic hydrocarbons	Trihalomethanes
Polynuclear aromatic hydrocarbons (PAH)	Chlorinated acetic acids
Chlorinated benzenes	Halogenated acetonitriles
Pesticides	

(THM). Among these THMs, four compounds are of importance, and guidelines values have been established (μg/L): bromoform (100), dibromochloromethane (100), bromodichloromethane (60), and chloroform (200) (WHO, 1984). Chlorophenols are another group and these cause odor problems even at low concentrations. Although measures should be taken to reduce formation of these disinfection by-products, disinfection should not be compromised, as bacteriological quality is most important in the supply of safe drinking water.

Drinking Water Quality Standards of Some Countries

The drinking water quality standards for Thailand and Indonesia are given in Tables 1.11(a)–(d). This is followed by a comparison in Table 1.11(e) of some important parameters in several developed and developing countries.

Table 1.11(a)　Standards for Drinking Water, Thailand (Laws and Standards on Pollution Control in Thailand, 1989)

			Standard values	
Properties	Parameters	Units	Max. acceptable concentration	Max. allowable concentration[a]
Physical	Color	Pt-Co	5	15
	Taste*	TTN	nonobjectionable (3)	nonobjectionable (3)
	Odor*	TON	nonobjectionable (3)	nonobjectionable (3)
	Turbidity	SSU(NTU)	5	20
Chemical	pH	—	6.5–8.5	≤9.2
	Total solids*	mg/dm^3	500 (600)	1500 (1000)
	Iron (Fe)	mg/dm^3	0.5	1.0
	Manganese (Mn)*	mg/dm^3	0.3 (0.1)	0.5 (0.3)
	Iron and Manganese (Fe and Mn)	mg/dm^3	0.5	1.0
	Copper (Cu)	mg/dm^3	1.0	1.5
	Zinc (Zn)	mg/dm^3	5.0	15
	Calcium (Ca)	mg/dm^3	75[b]	200
	Magnesium (Mg)	mg/dm^3	50	150
	Sulfate (SO_4)	mg/dm^3	200	250[c]
	Chloride (Cl)*	mg/dm^3	250	600 (500)
	Fluoride (F)	mg/dm^3	0.7	1.0
	Nitrate (NO_3)*	mg/dm^3	45 (10)	45 (10)
	Alkyl benzyl Sulfanates (ABS)	mg/dm^3	0.5	1.0
Toxic elements	Phenolic substance (as phenol)*	mg/dm^3 mg/dm^3	0.001	1.0 (0.002)
	Mercury (Hg)	mg/dm^3	0.001	
	Lead (Pb)	mg/dm^3	0.05	
	Arsenic (As)	mg/dm^3	0.05	
	Selenium (Se)	mg/dm^3	0.01	
	Chromium (Cr VI)	mg/dm^3	0.05	
	Cyanide (CN)*	mg/dm^3	0.2 (0.1)	

Table 1.11(a) Continued

			Standard values	
Properties	**Parameters**	**Units**	**Max. acceptable concentration**	**Max. allowable concentration**[a]
	Cadmium (Cd)*	mg/dm^3	0.01 (0.005)	
	Barium (Ba)	mg/dm^3	1.0	
Bacterial	Standard plate count	colonies/cm^3	500	
	Total coliform	MPN/100 cm^3	<2.2	
	E. coli	MPN/100 cm^3	none	
Epidemic	*Staphylococcus aureus***	MPN/100 cm^3	(none)	
bacteria	*Salmonella***	MPN/100 cm^3	(none)	
	*Clostridium perfringens***		(none)	
Pesticide	DDT (total)**	µg/dm^3	(1)	
residues	Aldrin and dieldrin**	µg/dm^3	(0.03)	
	Chlordane (total)**	µg/dm^3	(0.3)	
	Hexachloro benzene**	µg/dm^3	(0.01)	
	Heptachlor and heptachlor epoxide**	µg/dm^3	(0.1)	
	α-HCH**	µg/dm^3	(3)	
Radioactivity	Methoxy chloride	µg/dm^3	(30)	
	2,4-D**	µg/dm^3	(100)	
	Gross α**	Bq/dm^3	(0.01)	
	Gross β**	Bq/dm^3	(1)	

Note: Pt-Co = platinum cobalt scale, TON = threshold odor number, TTN = taste threshold number, SSU = silica scale unit, NTU = Nephelometric turbidity unit, MPN = most probable number, ( ) = proposed values.

[a]These values are allowed for tap water or groundwater that has to be used temporarily as drinking water. The characteristic is between maximum acceptable concentration and maximum allowable concentration and cannot be certified as standard drinking water.

[b]If calcium concentration is higher than the standard value and magnesium concentration is lower than the standard value, the Ca and Mg will be considered as total hardness, which should be less than 300 mg/dm^3 as CaCO$_3$.

[c]If the sulfate concentration reaches 250 mg/dm^3, magnesium concentration must not be higher than 30 mg/dm^3.

*Being revised by the Technical Committee Group 5 of Thai Industrial Standards Institute.

**New parameters being considered by the Technical Committee Group 5 of Thai Industrial Standards Institute.

Table 1.11(b) Bottled Drinking Water Quality Standards, Thailand
(Laws & Standard on Pollution Control in Thailand, 1989)

Properties	**Parameters**	**Units**	**Standard values (max. allowance)**
Physical	Color	Hazen	20
	Odor	—	none
	Turbidity	Silica scale	5.0
	pH	—	6.5–8.5

Table 1.11(b) Continued

Properties	Parameters	Units	Standard values (max. allowance)
Chemical	Total solids	mg/L	500
	Total hardness as $CaCO^3$	mg/L	100
	Arsenic (As)	mg/L	0.05
	Barium (Ba)	mg/L	1.0
	Cadmium (Cd)	mg/L	0.01
	Chloride as chlorine	mg/L	250
	Chromium (Cr)	mg/L	0.05
	Copper (Cu)	mg/L	1.0
	Iron (Fe)	mg/L	0.5
	Lead (Pb)	mg/L	0.1
	Manganese (Mn)	mg/L	0.05
	Mercury (Hg)	mg/L	0.002
	Nitrate as nitrogen (NO_3-N)	mg/L	4.0
	Phenol	mg/L	0.001
	Selenium (Se)	mg/L	0.01
	Silver (Ag)	mg/L	0.05
	Sulfate	mg/L	250
	Zinc (Zn)	mg/L	5.0
	Fluoride as fluorine (F)	mg/L	1.5
Bacterial	Coliform	MPN/100 ml	2.2
	E. coli	MPN/100 ml	none
	Disease-causing bacteria	MPN/100 ml	none

Table 1.11(c) Criteria of Water Quality Category A, Indonesia, 1990 (ADMAL, 1990)

No.	Parameter	Unit	Max. concentration	Notes
Physical				
1.	Odor	—	—	Odorless
2.	Total dissolved solid substances (TDS)	mg/L	1000	
3.	Turbidity	NTU scale	5	
4.	Taste	—	—	Tasteless
5.	Temperature	°C	Air temperature ±3°C	
6.	Color	TCU scale	15	
Chemical				
a. Inorganic Chemicals				
1.	Mercury	mg/L	0.001	
2.	Aluminum	mg/L	0.2	
3.	Arsenic	mg/L	0.05	
4.	Barium	mg/L	1.0	
5.	Iron	mg/L	0.3	
6.	Fluoride	mg/L	0.5	
7.	Cadmium	mg/L	0.005	
8.	$CaCO_3$ hardness	mg/L	500	
9.	Chloride	mg/L	250	
10.	Chromium (hexavalent)	mg/L	0.05	
11.	Manganese	mg/L	0.1	
12.	Sodium	mg/L	200	

Table 1.11(c)   Continued

No.	Parameter	Unit	Max. concentration	Notes
13.	Nitrate, as N	mg/L	10	
14.	Nitrite, as N	mg/L	1.0	
15.	Silver	mg/L	0.05	
16.	pH		6.5–8.5	Minimum and maximum limits
17.	Selenium	mg/L	0.01	
18.	Zinc	mg/L	5	
19.	Cyanide	mg/L	0.1	
20.	Sulfate	mg/L	400	
21.	Sulfide as H_2S	mg/L	0.05	
22.	Copper	mg/L	1.0	
23.	Lead	mg/L	0.05	

b. Organic Chemicals

No.	Parameter	Unit	Max. concentration
1.	Aldrin and dieldrin	mg/L	0.0007
2.	Benzene	mg/L	0.01
3.	Benzo (a) pyrene	mg/L	0.00001
4.	Chlordane (total isomer)	mg/L	0.0003
5.	Chloroform	mg/L	0.03
6.	2,4-Dichlorophenoxyacetic acid	mg/L	0.10
7.	DDT	mg/L	0.03
8.	Detergent	mg/L	0.5
9.	1,2-Dichloroethane	mg/L	0.01
10.	1,1-Dichloroethane	mg/L	0.0003
11.	Heptachlor and heptachlorepoxide	mg/L	0.003
12.	Hexachlorobenzene	mg/L	0.00001
13.	Lindane	mg/L	0.004
14.	Methoxychlor	mg/L	0.03
15.	Pentachlorophenol	mg/L	0.01
16.	Total pesticide	mg/L	0.1
17.	2,4,6-Trichlorophenol	mg/L	0.01
18.	Organic substances ($KMnO_4$)	mg/L	10

Microbiological

No.	Parameter	Unit	Max. concentration
1.	Fecal coliform bacteria	Total per 100 ml	0
2.	Total coliform bacteria	Total per 100 ml	3

Radioactivity

No.	Parameter	Unit	Max. concentration
1.	Gross alpha activity	Bq/L	0.1
2.	Gross beta activity	Bq/L	1.0

Note: NTU = nephelometric turbidity units, TCU = true color units. Heavy metals are as dissolved metals.

Table 1.11(d)   Criteria of Water Quality Category B, Indonesia (ADMAL, 1990)

No.	Parameter	Unit	Max. concentration
Physical			
1.	Temperature	°C	Normal water temp.
2.	Dissolved solid substances	mg/L	1000

Table 1.11(d) Continued

No.	Parameter	Unit	Max. concentration

Chemical

a. Inorganic Chemicals

No.	Parameter	Unit	Max. concentration
1.	Mercury	mg/L	0.001
2.	Free ammonia	mg/L	0.5
3.	Arsenic	mg/L	0.05
4.	Barium	mg/L	1
5.	Iron	mg/L	5
6.	Fluoride	mg/L	1.5
7.	Cadmium	mg/L	0.01
8.	Chloride	mg/L	600
9.	Chromium (hexavalent)	mg/L	0.05
10.	Manganese	mg/L	0.5
11.	Nitrate, as N	mg/L	10
12.	Nitrite, as N	mg/L	1
13.	Dissolved oxygen (DO)	mg/L	a
14.	pH		5–9
15.	Selenium	mg/L	0.01
16.	Zinc	mg/L	5
17.	Cyanide	mg/L	0.1
18.	Sulfate	mg/L	400
19.	Sulfide, as H_2S	mg/L	0.1
20.	Copper	mg/L	1
21.	Lead	mg/L	0.1

b. Organic Chemicals

No.	Parameter	Unit	Max. concentration
1.	Aldrin and dieldrin	mg/L	0.017
2.	Chlordane	mg/L	0.003
3.	DDT	mg/L	0.042
4.	Endrin	mg/L	0.001
5.	Phenol	mg/L	0.002
6.	Heptachlor and heptachlor epoxide	mg/L	0.018
7.	Carbon chloroform extract	mg/L	0.5
8.	Lindane	mg/L	0.056
9.	Methoxychlor	mg/L	0.035
10.	Oil and grease	mg/L	nil
11.	Organophosphate and carbamate	mg/L	0.1
12.	PCB	mg/L	nil
13.	Methylene blue active substance (surfactant)	mg/L	0.5
14.	Toxaphene	mg/L	0.005

Microbiological

No.	Parameter	Unit	Max. concentration
1.	Fecal coliform bacteria	Total per 100 ml	2,000
2.	Total coliform bacteria	Total per 100 ml	10,000

Radioactivity

No.	Parameter	Unit	Max. concentration
1.	Gross alpha activity	Bq/L	0.1
2.	Gross beta activity	Bq/L	1.0

Note: Heavy metals are as dissolved metals.

[a]Surface water is recommended to be higher than or at least 6.

Table 1.11(e) Comparison of Chemical and Physical Drinking Water Standards Recommended by the WHO and Several Developed and Developing Countries (Adapted from Schulz and Okun, 1984)

Chemical and physical parameters	WHO guideline values[a]	United States (1992)[b]	China (1976)[c]	Canada (1989)[d]	Japan (1993)[e]	Korea	Qatar	Tanzania (temporary) (1974)	Thailand (1989)[f]	Indonesia (1990)[g]	EC (1992)[h]
Total hardness (mg/L as $CaCO_3$)	—				300	300		600	300	500	
Turbidity (NTU)	5	0.5–1	5	1	2	2	5	30	5	5	1–10 SiO_2
Color (TCU)	15	15	15	15	5	2	20	50	15	15	20
Iron, as Fe (mg/L)	0.3	0.3	0.3	0.3	0.3	0.3	0.3	1	0.5	0.3	0.2
Manganese, as Mn (mg/L)	0.05	0.05	0.1	0.05	0.05	0.3	0.3	0.5	0.3	0.1	0.2
pH	<8	6.5–8.5	6.5–8.5	6.5–8.5	5.8–8.6			6.5–9.2	6.5–8.5	6.5–8.5	6.2–8.5
Nitrate, as NO_3 (mg/L)	50	45		45		45		100	45	10 mg/L as N	50
Sulfate, as SO_4 (mg/L)	250	400		500		200	250	600	250	400	250
Fluoride, as F (mg/L)	1.5	2	0.5–1.0	1.5	0.8	1.0	1.6	8.0	1	0.5	1.5
Chloride, as Cl (mg/L)	250	250		250	200	150	250	800	250	250	25
Arsenic, as As (mg/L)	0.01	0.05	0.04	0.025	0.01	0.05		0.05	0.05	0.05	0.05
Cadmium, as Cd (mg/L)	0.003	0.01	0.01	0.005	0.01		0.05	0.05	0.01	0.005	0.005
Chromium (mg/L)	0.05	0.05	0.05		0.01	0.05		0.05	0.05	0.05	0.05
Cyanide, as Cn (mg/L)	0.07	0.2	0.05	0.2	0.01			0.2	0.2	0.1	0.05
Copper, as Cu (mg/L)	2.0	1.0	1.0	1.0	1.0	1.0	0.3	3.0	1.0	1.0	3
Lead, as Pb (mg/L)	0.01	0	0.1	0.01	0.05	0.1	0.1	0.1	0.05	0.05	0.05
Mercury, as Hg (mg/L)	0.001	0.002	0.001	0.001	0.005				0.001	0.001	0.001
Selenium, Se (mg/L)	0.01	0.05	0.01		0.01			0.05	0.01	0.01	0.01

[a]WHO, 1993. [b]Pontius, 1992. [c]IDRC, 1981. [d]Decker and Long, 1992. [e]Japanese Standard, 1993. [f]National Environmental Board, 1989. [g]ADMAL, 1990. [h]AQUA, 1992.

1.3.2 Microbiological Aspects

The most important parameter of drinking water quality is the bacteriological quality, i.e., the content of bacteria and viruses. Ideally, drinking water should not contain any microorganisms known to be pathogenic. Microbial pathogens that may be found in water include salmonellas, shigellas, pathogenic vibrios, enteroviruses, diagnostic forms of pathogenic protozoa such as cysts of *Giardia lamblia*, as well as certain opportunistic (bacterial) pathogens (NHMRC-AWRC, 1987) (Table 1.12). Cyano-bacteria (blue-green algae) may produce toxins, and their development in drinking water supply impoundments should be prevented.

Table 1.12 Waterborne Disease-Causing Organisms (AWWA, 1990)

Name of organism or group	Major disease	Major reservoirs and primary sources
Bacteria		
Salmonella typhi	Typhoid fever	Human feces
Salmonella paratyphi	Paratyphoid fever	Human feces
Other salmonella	Salmonellosis	Human and animal feces
Shigella	Bacillary dysentery	Human feces
Vibrio cholera	Cholera	Human feces
Enteropathogenic *E. coli*	Gastroenteritis	Human feces
Yersinia enterocolitica	Gastroenteritis	Human and animal feces
Campylobacter jejuni	Gastroenteritis	Human and animal feces
Legionella pneumophila and related bacteria	Acute respiratory illness (legionellosis)	Thermally enriched waters
Mycobacterium tuberculosis	Tuberculosis	Human respiratory exudates
Other (atypical) mycobacteria	Pulmonary illness	Soil and water
Opportunistic bacteria	Variable	Natural waters
Enteric Viruses		
Enteroviruses		
Polioviruses	Poliomyelitis	Human feces
Coxsackieviruses A	Aseptic meningitis	Human feces
Coxsackieviruses B	Aseptic meningitis	Human feces
Echoviruses	Aseptic meningitis	Human feces
Other enteroviruses	Encephalitis	Human feces
Reoviruses	Mild upper respiratory and gastrointestinal illness	Human and animal feces
Rotaviruses	Gastroenteritis	Human feces
Adenoviruses	Upper respiratory and gastrointestinal illness	Human feces
Hepatitis A virus	Infectious hepatitis	Human feces
Norwalk and related GI viruses	Gastroenteritis	Human feces
Protozoans		
Acanthamocba castellani	Amoebic meningoencephalitis	Soil and water
Balantidium coli	Balantidosis (dysentery)	Human feces
Cryptosporidium	Cryptosporidiosis	Human and animal feces
Entamoeba histolytica	Amoebic dysentery	Human feces
Giardia lamblia	Giardiasis (gastroenteritis)	Human and animal feces

Table 1.12   Continued

Name of organism or group	Major disease	Major reservoirs and primary sources
Naegleria fowleri	Primary amoebic meningoencephalitis	Soil and water
Algae (blue-green)		
Anabaena flosaquae	Gastroenteritis	Natural waters
Microcystis aeruginosa	Gastroenteritis	Natural waters

Indicator Organisms

It is not practicable to test the water for all organisms that it might possibly contain. Instead, the water is examined for a specific type of bacteria that originates in large numbers from human and animal excreta and whose presence in the water is indicative of fecal contamination. Such indicative bacteria must be specifically fecal and not free-living. An indicator organism should always:

- be present when the pathogenic organism of concern is present, and absent in clean, uncontaminated water
- be present in fecal material in large numbers
- be able to respond to natural environmental conditions and to treatment processes in a manner similar to the pathogens of interest
- be easy to isolate, identify, and enumerate
- have a high ratio of indicator/pathogen
- come from the same source as the pathogen

Total coliforms, fecal coliforms, *E. coli*, fecal streptococci, enterococci and heterotrophic plate count (HPC) are some of the microorganisms selected to be indicators. However, total coliforms are the most commonly used indicator. Fecal coliforms and HPC also are used frequently. There are two methods for conducting tests on the levels of fecal coli and fecal streptococci in water: the multiple tube method for establishing the most probable number (MPN), and the membrane filtration method. Where possible, examination for both fecal coli and fecal streptococci should be made. This will provide an important check on the validity of the results. It will also give a basis for computing the ratio at which the two species of bacteria are present. From this, a tentative conclusion can be obtained whether the fecal pollution is of animal or human origin.

Total coliforms are defined as all aerobic and facultatively anaerobic, gram-negative, nonspore-forming, rod-shaped bacteria that ferment lactose with gas formation within 48 hours at 35°C. The definition includes *E. coli* plus species of enterobacter, klebsiella, and citrobacter. Fecal coliforms are a subgroup of the total coliforms. They can be distinguished in the laboratory through elevated temperature tests (43 to 45°C). HPC includes a broad group of bacteria including nonpathogens, pathogens, and opportunistic pathogens. HPC cannot be correlated with any likelihood of waterborne disease outbreak because of its lack of specificity. It is significant as a general indicator of poor biological quality.

Factors Increasing Bacterial Numbers

The factors that contribute to increase in the number of bacteria within a distribution may include the organic and microbiological load of the water, sediments, and biofilms, system parameters (such as loss of residual chlorine), and physical parameters such as temperature. Only a small percentage of the total organic load is available as organic matter for utilization by microorganisms within the aqueous phase. The remaining large macromolecules provide a conditioning film and concentrated nutrient source for subsequent biofilm development. High levels of organic matter affect disinfection creating a high chlorine demand and potential for formation of trihalomethanes (THM).

High turbidity levels are usually associated with increased number of microorganisms. Turbidity in a water source might be made up of suspended matter such as clay, silt, inorganic and organic matter, plankton, and microscopic organisms. This turbidity will add to sediments in areas of slow flows and dead ends in the treatment and distribution systems. The presence of turbidity also decreases the chlorination efficiency.

A biofilm is an organic or inorganic surface deposit consisting of microorganisms, microbial products, and detritus. Biofilms comprise microorganisms and extracellular products. In a distribution or transmission network, the constant removal of biofilm from the pipe surface back into the aqueous phase will contribute to the deterioration of drinking water quality.

Figure 1.3 shows the four main stages of biofilm development. As the total prevention of biofilm development in water supply is not practicable, the best option available at present is the minimization of biofilm accumulation.

The standards for microbiological quality are presented in Tables 1.13(a) and 1.13(b) (in summary form).

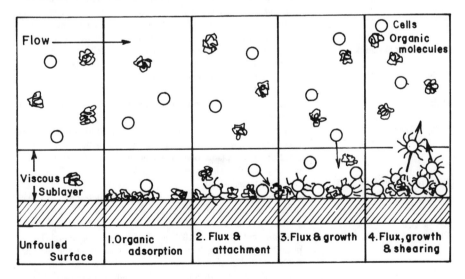

FIGURE 1.3   The stages of biofilm development.

Table 1.13(a)   Standards for Microbiological Quality
(Adapted from Sayre, 1988; WHO, 1993 and NHMRC-AWRC, 1987)

Characteristic	U.S.	Canadian	EEC	WHO	Australian guideline
Fecal coliform (organisms per 100 ml)	0	0	0	0	0
Coliforms (organisms per 100 ml)	1	10	—	0	0 (95% of samples analyzed), remaining <10 per 100 ml
Total bacteria count (supplied for human consumption)					
37°C	—	—	10 per ml	—	—
22°C	—	—	10 per ml	—	—

Note: — = not specified.

Table 1.13(b)   Comparison of Microbiological Drinking Water Standards

WHO guideline values[a]	1. All water intended for drinking; *E. coli* or thermotolerant coliform bacteria 0 per 100 ml
	2. Treated water entering the distribution system; *E. coli* or thermotolerant coliform bacteria, total coliform bacteria 0 per 100 ml
	3. Treated water in the distribution system; *E. coli* or thermotolerant coliform bacteria, total coliform bacteria 0 per 100 ml. For large supplies, where sufficient samples are examined, must not be present in 95% of samples taken throughout any 12 month period.
United States[b]	Number of coliform bacteria as determined by membrane filter test shall not exceed 1 per 100 ml as the arithmetic mean of all samples examined per month. When 10 ml fermentation tubes are used, coliform bacteria shall not be present in more than 10% of the portions in any month. When 100 ml tubes are used, coliform shall not be present in more than 60% of the portions in any month.
European Community (1992)[c]	The number of total coliforms, fecal coliforms and fecal streptococci determined by the membrane filter method should be 0. When determined by multiple tube method, the multiple tube number (MPN) should be less than 1.
China[d]	Total colony count not more than 100 per ml; *E. coli* not more than 3 per ml.
India (1973)[e]	Coliform = 0 to 1.0 per 100 ml permissive; 10 to 100 per 100 ml excessive but tolerable in absence of alternative, better source; 8 to 10 per 100 ml acceptable only if not in successive samples; 10% of monthly samples can exceed 1 per 100 ml.
India recommended (1975)[e]	*E. coli* = 0 per 100 ml. Coliform = 10 per 100 in any sample, but not detectable in 100 ml of any two consecutive samples or more than 50% of samples collected for the year.
Philippines (1963)[e]	Coliform: not more than 10% of 10 ml portions examined shall be positive in any month. Three or more positive 10 ml portions shall not be allowed in two consecutive samples; in more than one sample per month when less than 20 samples examined; or in more than 5% of the samples when 20 are examined per month.

Table 1.13(b) Continued

Qatar[e]	Coliforms = 0 per 100 ml if present in two successive 100 ml samples is considered as grounds for rejection of supply.
Tanzania (temporary 1974)[f]	Nonchlorinated pipe supplies: 0 per 100 ml coliform, classified as excellent; 1 to 3 per 100 ml coliform, classified as satisfactory; 4 to 10 per 100 ml coliform, classified as suspicious; 10 per 100 ml coliform, classified as unsatisfactory; one or more *E. coli* per 100 ml classified as unsatisfactory. Other supplies: WHO standards to be aimed at.
Thailand[e]	Coliform = 2.2 per 100 ml. *E. coli* = 0 per 100 ml.

Note: The data before 1980 are presented to enable comparison with the current standards.

[a]WHO, 1993. [b]Pontius, 1992. [c]AQUA, 1992. [d]IDRC, 1981. [e]Schulz and Okun, 1984. [f]National Environmental Board, 1989.

Table 1.14 presents WHO guidelines for bacteriological quality for treatment prior to and in the distribution system. Other details can be found in the literature (WHO, 1993).

Regrowth of Microorganisms in Distribution Systems

Many factors may contribute to the regrowth of bacteria in drinking water. The loss of residual disinfectant or bacterial breakthrough are significant causes. Distribution systems unable to maintain an effective disinfectant may experience growth of HPC and coliform bacteria. McCabe et al. (1970) showed that chlorine residual levels of 0.2 mg/L or more were associated with HPC levels of < 500 cfu/ml in 98% of water samples. Resolving a regrowth problem in this situation is simple. The system can be flushed and disinfectant applied, ensuring that the residual level is maintained in all parts of the system. In some cases, rechlorination facilities may be used to increase the residual level. The removal of demand-causing compounds through selection of appropriate treatment, pipe relining, or mains replacement may help in preventing regrowth of microorganisms.

Breakthrough is the increase in bacteria in the distribution system resulting from inadequate water treatment. Breakthrough of coliform organisms in treatment plants may even occur when samples are apparently of good microbiological quality. Injured bacteria may result from incomplete disinfection and may not be detected using standard coliform media. Recent research indicated that injured coliform bacteria can repair the cellular lesion and resuscitate in biofilms (Wales et al., 1989). The solution to breakthrough and subsequent regrowth is to eliminate the source of contamination. Application of the sensitive microbiological media m-T7 agar in sampling will help detect injured coliform bacteria.

Sampling Frequency

In piped water supplies, the probability of contamination of the distribution system increases with the length of pipe network and the number of plumbing systems attached to it. In assessing the quality of water and effectiveness of disinfection, although it is desirable to take samples at least weekly, this may not be possible with

Table 1.14   WHO Guideline Values for Bacteriological Quality

Organisms	Guideline value
All water intended for drinking	
E. coli thermotolerant coliform bacteria[a,b,c]	Must not be detectable in any 100-ml sample
Treated water entering the distribution system	
E. coli thermotolerant coliform bacteria[b]	Must not be detectable in any 100-ml sample
Total coliform bacteria[a]	Must not be detectable in any 100-ml sample
Treated water in the distribution system	
E. coli thermotolerant coliform bacteria[b]	Must not be detectable in any 100-ml sample
Total coliform bacteria	Must not be detectable in any 100-ml sample. In the case of large supplies, where sufficient samples are examined, must not be present in 95% of samples taken throughout any 12-month period.

Reproduced with permission from *Guidelines for Drinking Water Quality*, WHO, 1993.

[a]Immediate investigative action must be taken if either *E. coli* or total coliform bacteria are detected. The minimum action in the case of total coliform bacteria is repeat sampling; if these bacteria are detected in the repeat sample, the cause must be determined by immediate further investigation.

[b]Although *E. coli* is the more precise indicator of fecal pollution, the count of thermotolerant coliform bacteria is an acceptable alternative. If necessary, proper confirmatory tests must be carried out. Total coliform bacteria are not acceptable indicators of the sanitary quality of rural water supplies, particularly in tropical areas where many bacteria of no sanitary significance occur in almost all untreated supplies.

[c]It is recognized that, in the great majority of rural water supplies in developing countries, fecal contamination is widespread. Under these conditions, the national surveillance agency should set medium-term targets for the progressive improvement of water supplies.

small systems. If possible, decisions on sampling frequency should be taken by national authorities. The following minimum sampling frequencies are recommended, the samples being spaced out evenly throughout the month (WHO, 1993).

Population served	Minimum number of samples
less than 5,000	1 sample per month
5,000–100,000	1 sample per 5,000 population per month
more than 100,000	1 sample per 10,000 population per month

EXAMPLE

Population Calculation

The population of a town as per census records are given below for the years 1929 to 1989. Assuming that the scheme of water supply will commence to function from 1994, it is required to estimate the population in 2024.

Year	Population
1929	40,185
1939	44,522
1949	60,395
1959	75,614
1969	98,886
1979	124,230
1989	158,800

Year	Population	Increment (X)	Incremental Increase (Y)	Rate of Growth
1929	40,185	—	—	—
1939	44,522	4,337	—	0.108
1949	60,395	15,873	+ 11,535	0.357
1959	75,614	15,219	− 654	0.252
1969	98,886	23,272	+ 8,053	0.308
1979	124,230	25,344	+ 2,072	0.256
1989	158,800	34,570	+ 9,226	0.278
		average: 19,769	average: 6,047	

Arithmetic Progression Method

Increase in population between 1929 and 1989 $= 158,800 - 40,185$
$$= 118,615$$

Average increase per decade $= \dfrac{1}{6} \times 118,615 = 19,769$

Population at the end of the design period (i.e., in 2024)
$\quad$ = population in 1989 + (increment × no. of decades)
$\quad$ = 158,800 + (19,769 × 3.5)
$\quad$ = 227,992

Geometric Progression Method

Geometric mean of growth rate (r)

$r = \sqrt[6]{0.108 \times 0.357 \times 0.252 \times 0.308 \times 0.256 \times 0.278}$
$\quad = 0.2443$

Assuming that the future growth follows the geometric mean of growth rate during the period 1929 to 1989,

Design population (in 2024) $= P_{1989} (1 + 0.2443)^{3.5}$
$\quad\quad\quad = 158,800 (1 + 0.2443)^{3.5}$
$\quad\quad\quad = 341,262$

Method of Varying Increment or Incremental Increase Method

Average incremental increase = 6047

Design population $P_{2024} = P_{1989} + (3.5 \times 19,769) + (3.5 \times 4.5 / 2) \times 6047$
$$= 275,611$$

Graphical Projection Method

From the regression analysis, one could establish an empirical equation and predict the future population.

REFERENCES

ADMAL, A guide to environmental assessment in Indonesia, government regulation of the Republic of Indonesia, number 20, 1990.

AQUA, Drinking water directive 80/778/EC, Bureau's views on proposals for modifications, *AQUA*, 41(2), 101–108, 1992.

AWWA, *A Hand Book of Community Water Supplies*, McGraw-Hill, 1990.

Carney, M., European drinking water standards, *J. AWWA*, 83, 48, 1991.

Chung, Y., *The risk assessment and management of drinking water*, Proceedings of Korea-Australia joint seminar on the recent trends in technology development for water quality conservation, Seoul, Korea, June 1993, 47–66.

Decker, K. C. and Long, B. W., Canada's cooperative approach to drinking water regulation, *J. AWWA*, 84(4), 120, 1992.

Design Manual for Water Supply and Treatment, India, Third Edition, Ministry of Urban Development, New Delhi, March 1991.

IDRC, *Rural Water Supply in China*, International Development Research Center, Ottawa, Canada, 1981.

IRC, Internal Reference Center for Water Supply and Sanitation, *Small Community Water Supplies*, Technical Paper Series 18, Rijswijk, The Netherlands, 1981.

Japanese Standard for Drinking Water Quality, Established 21 Dec 1992 by the Director General of Water Supply and Environmental Sanitation Department, and enforced 1 Dec 1993.

Jolliffe, I., *Notes on Water Supply Systems*, prepared for the subject on Water Supply and Sewerage Systems, University of Technology — Sydney, Lectures, 1991.

Laws and Standards for Pollution Control in Thailand, Second Edition, Environmental Quality Standards Division, Office of the National Environmental Board, Thailand, July 1989.

McCabe et al., Study of community water supply systems, *J. AWWA*, 62(11), 670, 1970.

NHMRC-AWRC, *Guidelines for Water Quality in Australia*, Canberra, Australia, 1987.

Pontius, F. W., A current look at the federal drinking water regulations, *J. AWWA*, 82(3), 36, 1992.

Sayre, I. M., International standards for drinking water, *J. AWWA*, 80, 53, 1988.

Schulz, C. R. and Okun, D. A., *Surface Water Treatment for Communities in Developing Countries*, 1984.

Wales, S., McCaters, G. A. and Chevellier, L. E., Reactivation of Injured Bacteria, *J. Environ. Microbiol.*, 1989.

WDR — World Development Report, *Clean Water and Sanitation, Development and the Environment*, Executive Summary, 5, 16–17, 1992.

WHO — World Health Organization, *Guidelines for Drinking Water Quality, Vols. 1 and 2*, Geneva, Switzerland, 1984/1993.

2: Water Sources and Treatment Technologies

CONTENTS

2.1 INTRODUCTION

Water is usually withdrawn for drinking and household purposes from the following sources:

- groundwater (springs, infiltration galleries, wells, etc.)
- surface water (streams, lakes, ponds, rivers, impounded reservoirs, stored rain water, etc.)

If groundwater is found in adequate quantity and if it is conveniently located for the water supply concerned, the groundwater should be used, as it is less polluted compared with surface water.

Groundwater may be aerobic or anaerobic, depending on the environmental conditions where it is found. The anaerobic groundwater often contains CO_2, which make it corrosive. The removal of CO_2 as well as provision of oxygen can be facilitated by aeration. Chlorination also removes CO_2. Some groundwater contains excessive amounts of Fe, Mn, hardness, or fluoride, and these should be removed.

The surface waters are generally more polluted than groundwater due to their exposure to the environment, hence they may require more treatment steps than groundwater. The typical impurities may include turbidity, color, algae, floating debris, bacteria and other microorganisms, etc., in addition to the constituents of groundwater.

Surface water in general contains physical, chemical, and biological impurities (such as clay, sand, colloids, minerals, color, odor, taste, microorganisms). Rivers and streams, because of the flowing nature of the water and exposure to sunlight, have self-purification properties. However, they carry sand, silt, and clay. The intake point for the water supply should be far from the sewage effluent discharge point. Lakes and ponds are generally contaminated with bacterial impurities because the water is stagnant. Water quality in impounded reservoirs is highly variable, depending on the season. Table 2.1 presents the common pollutants in different water sources and the essential and optional treatment methods available for their removal.

2.2 CONVENTIONAL WATER TREATMENT PROCESSES

The most common water treatment processes used for a treatment of raw water from a surface source are rapid mixing, flocculation, sedimentation, filtration, and disinfection. In addition, there are treatment methods specific to some impurities such as iron, manganese, and fluorides. These impurities are mainly associated with groundwater, as opposed to surface water supplies. Only those impurities that are found to cause significant difficulties have been considered for specific treatment. For example, in some water supplies, fluorides are found in excess and therefore a specific treatment technique has to be applied to bring fluoride levels within allowable limits. Even though there are several alternative methods available for each treatment process (Table 2.2), only the simple techniques are reviewed in this book.

Rapid mixing is commonly used to disperse the chemicals added to water so that the chemicals come into close contact with the impurities in water. Rapid mixing is also used in disinfection to mix the chlorine uniformly with the mass of water. The common method of rapid mixing is mechanical mixing. This, however, consumes considerable energy and requires a high degree of maintenance. Due to these disadvantages, nonmechanical or hydraulic devices for mixing such as hydraulic jump are gaining popularity. In some cases, in-line blenders and injection mixers have been found to be technically sound and cost-effective alternatives. They are discussed in detail in Chapter 3.

Flocculation follows rapid mixing and facilitates the agglomeration of particles and minute flocs into large flocs by gradually bringing the particles together. The mixing intensity (velocity gradient) plays an important role in flocculation unlike in rapid mixing and therefore, the extent of agitation must be very carefully controlled. Flocculation is generally achieved through slow mixing, again using mechanical devices. The process is characterized by relatively high energy consumption and operation and maintenance costs. To overcome these disadvantages, hydraulic methods (such as gravel bed flocculator and Alabama-type flocculators) that make use of

Table 2.1 Selection of Water Treatment Processes Depending on the Common Impurities Present in Raw Water

Pollutant	Essential treatment steps	Optional treatment steps
Floating matter	Screening	Microstraining
Algae	Coagulation, flocculation and sedimentation, followed by rapid sand filtration	Microstraining, prechlorination, addition of inhibiting chemicals (e.g., $CuSO_4$)
Turbidity (NTU)		
<5	Slow sand filtration and postchlorination	Sedimentation and rapid filtration
<5–100	Coagulation, sedimentation, rapid filtration, and postchlorination	Flocculation and slow sand filtration
100–750	Coagulation, sedimentation, rapid filtration, and postchlorination	Screening, preliminary settlement, flocculation, slow sand filtration
750–1000	Coagulation, sedimentation, rapid filtration, and postchlorination	Raw water storage, screening, preliminary settlement, flocculation
>1000	Preliminary settlement, coagulation, sedimentation, rapid filtration and postchlorination	Raw water storage, screening
Color (if greater than 30 Hazen units)	Coagulation, flocculation, sedimentation, and RSF	
Tastes and odors		Aeration, prechlorination, slow sand filtration, adsorption by activated carbon
Hardness (>200 mg/L)	Lime softening	Lime/soda ash treatment

Table 2.1 Continued

Pollutant	Essential treatment steps	Optional treatment steps
Iron and manganese		
<0.3 mg/L		Aeration and prechlorination
0.3–1 mg/L	Coagulation, sedimentation, and rapid sand filtration	Aeration and slow sand filtration
>1 mg/L	Aeration, prechlorination, coagulation, sedimentation, and rapid filtration	
Fluorides (>1 mg/L)	Alum flocculation, sedimentation, lime softening, or ion exchange	
Chlorides (if >500 mg/L)	Desalting	
Coliform bacteria MPN/100 mL		
0–20	Postchlorination	
20–100	Coagulation, sedimentation, and postchlorination	Slow sand filtration
100–5000	Coagulation, sedimentation, and postchlorination	Prechlorination and slow sand filtration
>5000	Preliminary sedimentation, coagulation, sedimentation, filtration, and postchlorination	Raw water storage and primary sedimentation
Toxic chemicals	Should be used only if alternative water sources are absent. Coagulation, sedimentation, RSF, and activated carbon treatment	

Table 2.2   Selected Unit Operations in Water Supply

Unit operations	Conventional	Modified process	Comments
Rapid mixing	Hydraulic mixers; mechanical backmix reactor	Flash mixing facility (injection type)	Suitable for water treatment plants of all sizes. Higher percentage of coagulant utilization (much better effluent quality).
Flocculation	Hydraulic and mechanical type flocculation	Tapered flocculation	Less energy input, better floc formation.
		Gravel bed flocculator	Simplicity, effective flocculation, suitable to be used in low-cost package plants.
		Alabama flocculator	Economical to construct, operate, and maintain; minimum supervision needed.
Sedimentation	Rectangular horizontal flow filtration	Solids contact clarifier	Smaller size, better efficiency.
		Tube settler	Modular design; surface loading rate is 2 to 10 times or greater, thus very compact; upgrading of existing treatment plants possible by placing modules in the existing sedimentation tank.
Filtration	Rapid sand filtration	Dual media filter Coarse media filter Declining rate filter	Better filtration efficiency, longer filter run.
		Direct filtration	No sedimentation; smaller flocculation facility, lower chemical use, low energy requirement.
Disinfection	Chlorine	ClO_2 O_3	No THM formation, lower chemical requirement, better disinfection.

gravitational energy are being more frequently used, particularly in small to medium-sized plants. These techniques are discussed in detail in Chapter 4, together with the compartmentalized tapered flocculation arrangement.

Subsequent to coagulation and flocculation, the flocs are allowed to settle in a settling tank. Conventionally, sedimentation has been effected in rectangular clarifiers or at times in circular clarifiers. There are sedimentation tanks with numerous modifications, but the basic principles are the same. The flocs formed generally require a long time to settle down and thus require large tanks to obtain satisfactory effluent.

High rate settling processes were always the dream of water supply engineers, and there are several alternatives available. These have never been in much demand in developing countries because of the simplicity of rectangular/circular basins and the availability of cheap land. But there are always situations where land is scarce, particularly in urban areas. There are other instances where the capacity of the plant must be increased to meet the increase in demand for water. Tube settlers are ideal in such situations. Combinations of more than one process are also feasible. An example is the sludge blanket clarifier or clariflocculator, where coagulation, flocculation, and clarification take place in the same tank. The design criteria for these techniques are well established and therefore their application is quite straightforward. They are discussed in Chapter 5.

Filtration is an age-old technique to capture the fine particles in water. Even bacteria and viruses can be trapped in a filter to a certain extent. Slow sand filtration has been extensively used in small community water supplies in developing countries due to its simplicity in design, operation, and maintenance. Rapid sand filtration, on the other hand, is used in urban water treatment plants. There have been a number of modifications made on the rapid sand filter to improve the filtrate quality and to meet the increasing demand for water supply. Both slow and rapid sand filters are discussed with their modifications and their applicability status in Chapter 6. Selected low-cost filtration systems are also presented in this chapter.

It always becomes necessary to treat raw water for removal of iron, manganese, fluorides, etc. The treatment techniques discussed so far are not effective in removing these impurities and, therefore, specific treatments are necessary. Technologies used to remove iron, manganese, and fluorides are discussed in Chapter 7.

No water treatment process is complete without disinfection. In fact, the most important of all the water treatment processes is disinfection because it prevents waterborne diseases in humans. Chlorine has been used extensively for decades as a disinfectant. But problems and after-effects due to chlorine use, such as trihalomethane (THM) formation, have encouraged the use of alternative disinfectants as well as alternative techniques of disinfection. Alternative disinfectants such as ozone and chlorine dioxide are reviewed in detail. In Chapter 8, simple dosing techniques, along with their specific applications are presented.

Conventional water treatment processes often do not remove dissolved organics and inorganics, algae, color, odor, and viruses in significant quantities. Flotation, adsorption, ion exchange, and membrane processes that can be used for their removal are discussed in the next two sections.

2.3 ADVANCED PROCESSES

2.3.1 Flotation

Flotation is used to separate solids or dispersed liquids from a liquid phase. The separation is effected by introducing fine gas bubbles, usually air, into the system. The added fine air bubbles either adhere to or are trapped in the particle's structure, making the particles buoyant and bringing them to the surface. Even particles with

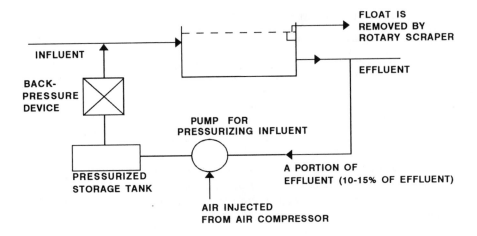

FIGURE 2.1 Flow diagram of dissolved air pressure flotation system.

a density greater than that of the liquid phase can be separated by flotation. Surface properties of particles play a predominant role during flotation rather than size or relative density.

The types of flotation in use are dispersed and dissolved air flotation. In dispersed air flotation, air is directly introduced into the liquid through diffusers whereas, in the case of dissolved air flotation, air bubbles are produced by precipitation from a solution supersaturated with air. Production of air bubbles can be achieved by dissolved air pressure flotation. Here the influent to the flotation unit is pressurized and then released in the unit to produce air bubbles.

Dissolved air pressure flotation is used in removing algae in water treatment plants. A typical flow diagram of a dissolved air pressure flotation system is shown in Figure 2.1. The important factors in designing the unit are the influent solids concentration, the quantity of air expressed as air-to-solid ratio, and the overflow rate. For better design of the flotation unit, laboratory test of preliminary design and pilot plant studies are always recommended.

2.3.2 Adsorption

Adsorption is used to remove taste, odor, color, organic impurities, nondegradable organics, heavy metals, etc., from raw water. In most cases, adsorption is used as a final treatment process prior to disinfection. The most commonly used adsorbent is activated carbon, but other materials like peat, wood, charcoal, fly ash, and slag are also in use.

Adsorption is a physical phenomena by which molecules of the solids are attached on the surface of adsorbent materials due to the intermolecular forces of attraction and are thereby removed. An adsorption unit can be operated either as fixed or fluidized bed. A fixed bed adsorption unit when operated at countercurrent mode will be efficient. A complete activated carbon adsorption system includes carbon

Table 2.3 Typical Design Values of a Fixed-Bed Adsorption System

Reactor diameter	0.6–3.6 m
Area loading	100–500 L/min/m²
Organic loading	0.1–0.3 kg BOD or COD/kg of carbon
Bed depth	1.5–9.0 m
Contact time	10–50 min
Backwash	500–1000 L/min/m²
Air scour	1.0–1.5 m³/min/m²
Spent carbon	3.0–10 kg/kg of COD removed for tertiary treatment

storage vessels and thermal regeneration facilities. During operation, the carbon must be disposed of or regenerated once its adsorptive capacity has been fully utilized. The regeneration of carbon can be done using a multiple-hearth furnace. The adsorption process is affected by the following factors:

- surface area of adsorbent
- nature of adsorbate
- pH
- temperature
- solute concentration
- time of contact
- nature of contacting system

Since many factors affect the process, it is not advisable to use generalized design criteria. It is always better to perform a laboratory or pilot-scale study with the raw water. Typical design values of a fixed bed adsorption system is presented in Table 2.3.

2.3.3 Ion Exchange

Ion exchange is a process of exchanging certain cations and anions in the water with sodium, hydrogen, or other ions held in the resinous ion exchange material. In industrial waste treatment, it is used to recover valuable waste materials as by-products, particularly ionic forms of precious metals such as silver, gold, and uranium.

In normal operation, the ion exchanger is in contact with the solution containing the ion to be removed until the active sites in the exchanger are partially or completely exhausted by that ion. The exchanger is then contacted with sufficiently concentrated solution of the ion originally associated with it to regenerate to its original form.

The exchange reactions may be written as follows:

$$
\left.\begin{matrix} Ca^{++} \\ \} \\ Mg^{++} \end{matrix}\right. \begin{matrix} (HCO_3)_2^- \\ SO_4^{--} \\ Cl_2^- \end{matrix} + Na_2^+ Z^- \rightarrow \left.\begin{matrix} Ca^{++} \\ \} \\ Mg^{++} \end{matrix}\right. Z^- + \begin{matrix} 2Na^+ HCO_3^- \\ Na_2^+ SO_4^{--} \\ 2Na^+ Cl^- \end{matrix}
$$

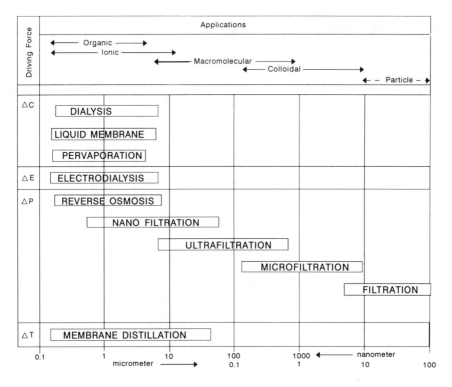

FIGURE 2.2 Effective range of membrane processes. ΔT = temperature, ΔP = pressure, ΔE = electric potential, ΔC = concentration.

and the regeneration reaction is as follows:

$$\left.\begin{array}{c} Ca^{++} \\ \\ Mg^{++} \end{array}\right\} \quad Z^- + \ 2Na^+Cl^- \quad \rightarrow \quad \left.\begin{array}{c} Ca^{++} \\ \\ Mg^{++} \end{array}\right\} \quad Cl_2^- + Na_2^+Z^-$$

where Z = zeolite radical.

The ion-exchange process can be used with cations and anions, both organic and inorganic. However, most of the applications of ion exchange involves inorganic species which often require the use of extremely highly concentrated regenerant and/ or the use of organic solvents to remove organic species. Different varieties of cation and anion exchangers are available for exchange with a particular ion of different stabilities. Generally, ions with higher charge will form more stable salts with the exchanger than those with lower charge; hence, polyvalent species can more frequently be removed from a solution than monovalent ones.

Three modes of ion-exchange operations used in industrial applications are concurrent fixed-bed, countercurrent fixed-bed, and continuous countercurrent.

2.4 MEMBRANE PROCESSES

Membrane processes used in water and wastewater treatment can be categorized into five classes according to the size of particles that can be retained, namely, reverse osmosis (RO), nanofiltration (NF), ultrafiltration (UF), microfiltration (MF), and electrodialysis (ED). ED is a proven process for desalting brackish water. RO is also used extensively in desalting applications. It has an added advantage of being able to remove many organic compounds in addition to ionic species and microorganisms. The NF, UF, and MF techniques are useful in removing macromolecules, colloids, and suspended solids.

Membrane processes are chosen according to the size range of solutes in the solution. Figure 2.2 gives the size ranges of some typical impurities found, and the applicable membrane processes.

2.4.1 Reverse Osmosis

Reverse osmosis is based on the well-known phenomenon of osmosis, which occurs when two solutions of different concentrations are separated by a semipermeable membrane. In this process, pressure is applied on the side of the concentrated solution to reverse the natural osmotic flow. The thin RO membranes are essentially nonporous, and they preferentially pass water and retain most solutes including ions. The rejection (or retention) of ions is typically in the range 95 to 99.9% depending on the ions and the membrane. RO is characterized by high operating pressures (20 to 100 bars).

The RO unit was first used for desalination of sea water. At present, there are large installations with a capacity of 2×10^6 m³/d to treat sea water for domestic and industrial purposes. In addition, there are small and medium-sized RO installations existing (0.4 to 95,000 m³/d) to supply pure water for specific purposes like petroleum platforms, agricultural purposes, sterilized water for hospitals, laboratories, etc. (Vigneswaran and Ben Aim, 1989).

In some countries, the surface water as well as water from aquifers contains nitrates in excess and therefore has to be treated. RO can be utilized for this purpose.

The RO process is used for the production of pure water for industrial purposes. One of the main industrial uses of RO is to prepare ultra-pure water for the electronic industry. RO membranes remove the contaminants except dissolved gases. RO units are also utilized to produce sterilized water for pharmaceutical industries and for medical purposes since they produce water absolutely free from bacteria and suspended solids. RO should be able to retain all species of concern, except for some organic species that may be partially transmitted by some types of RO membrane.

However, RO systems have some limitations. The operating range of RO falls in the order of a nanometer. The rate of permeate flux is very low though the pressure applied is as high as 20–100 bars, whereas other membrane processes operate at comparatively lower pressure and a higher flux rate. Because of low flux, RO needs a larger membrane surface area. Furthermore, RO does not operate successfully for higher solute removal with solutions of high concentration. For optimum performance, a good pretreatment should be provided. In addition, due to concentration polarization, gel layer formation, fouling, and internal clogging, the permeate flux rate decreases as the membrane filtration proceeds.

2.4.2 Nanofiltration

The nanofiltration membranes are believed to have pores (2 to 5 nm) and partially retain ions, though small and monovalent ions and low molecular weight organics tend to pass. For example NF membranes with 50% retention of NaCl and 98% retention of $MgSO_4$ are available. They can satisfactorily remove viruses and natural organics found in water. NF membranes usually have significantly higher water permeability than RO membranes and operate at lower pressure (typically 7 to 30 bars).

NF membranes should be capable of removing most contaminants including aluminum in its various forms. As a free multivalent ion, it should be retained by a suitable NF membrane. As a complex polyion, it should be large enough to be retained.

2.4.3 Ultrafiltration

Ultrafiltration membranes allow the passage of water and retain high molecular weight solutes and colloidal particulates. UF pore sizes usually range from 5 to about 20 nm, and retain fine colloids, macromolecules, and microorganisms. Partial retention of some ions may occur due to the charge interaction. UF membranes typically operate with pressures in the range of 1 to 10 bars.

UF is widely used to produce sterilized water in the pharmaceutical industry and for medical applications. In closed-circuit aquaculture practices, UF can be used as a sterilizing unit instead of an UV sterilizer.

The major advantages of UF are that it operates at lower pressure and yields a higher permeate flux compared to RO. However, microfiltration (MF) operates at a still higher permeate flux, 10–100 times higher than that of UF, with a lower pressure in the order of 1–2 bars.

2.4.4 Microfiltration

Microfiltration is an important separation process, as the permeate flux is higher than that of any other membrane processes and the permeate quality is much better than that of the conventional separation processes such as sedimentation, centrifugation, filtration, flotation, etc. Furthermore, most of the pollutants present in water have particle sizes ranging from 0.05 to 10 μm and can be removed by MF as they fall within the range of the microfiltration process. It can retain colloids, microorganisms, and suspended solids. MF membranes can often retain species much smaller than their rated pore size due to adsorptive capture. MF membranes typically operate at 0.5 to 5 bars pressure. Table 2.4 summarizes the typical differences between the various pressure-driven membrane processes.

Table 2.4   Differences Between the Various Pressure Driven Membrane Processes

	MF	UF	NF	RO
Membrane	Porous isotropic	Porous asymmetric	Finely porous asymmetric/composite	Nonporous asymmetric/composite
Pore size	50 nm–10 µm	5–20 nm	2–5 nm	—
Transfer mechanism	Sieving and adsorptive mechanisms (solutes migrate by convection)	Sieving and preferential adsorption	Sieving/electrostatic hydration/diffusion	Diffusion (solutes migrate by diffusion mechanisms)
Law governing transfer	Darcy's Law	Darcy's Law	Fick's Law	Fick's Law
Typical solution treated	Solution with particles	Solution with colloids and/or macromolecules	Solution with ions, small molecules	Solution with ions, small molecules
Typical pure water flux (L/m²·d)	500–10,000	100–2,000	20–200	10–100
Pressure applied (bar)	0.5–5	1–10	7–30	20–100

2.4.5 Hybrid Processes

Hybrid processes in which MF (or UF) is coupled with a chemical coagulant or powered activated carbon (PAC) in a slurry provide improved removal efficiency compared with the basic membrane processes.

Table 2.5 lists typical characteristics of raw water and the treated water objectives for membrane processes. The protozoa *Cryptosporidia* and *Giardia*, which as cysts or oocysts are quite difficult to remove by conventional filtration and are resistant to chlorination, can successfully be removed by membrane processes. Table 2.5 shows that removal efficiencies of 100% for microorganisms, up to 97% for turbidity, up to 90% for color, and 80% to 90% for certain key organics are essential.

All the membrane processes are capable of almost complete removal of bacteria. The degree of removal of virus by MF is a moot point. While the micropores may be typically an order of magnitude larger than a virus, substantial reductions are feasible. There are two reasons for this. First, waterborne viruses are usually associated with particulate or colloidal material that is removable by microfiltration. Second, viruses would tend to be adsorbed to the internal surface of the membrane, particularly if the membrane is hydrophobic (similar observations have been made with the filtration of protein solutions, and the explanation is attributed to membrane-species interactions, either hydrophobic or electrostatic). It is expected that improved removals would be achieved with the addition of chemicals.

Table 2.5 Raw and Treated Water Quality and Required Removal by Membrane Processes

Parameter	Raw water (min–max)	Treated water objectives	Removal required (% of maximum)
Turbidity (NTU)	0.55–100	<0.3	>97
Color (Hazen)	7–48	<5	>90
Iron (mg/L)	0.11–1.12	<0.2	>82
Mn (mg/L)	0.002–0.370	<0.03	>92
Al (mg/L)	0.003–1.40	<0.2	>87
Hardness (mg/L)	5.2–23.0	—	—
Alkalinity (mg/L)	0.6–7.6	—	—
Total coliform (numbers/100 ml)	Very wide variation	0	100
Fecal coliform (numbers/100 ml)	depending on the	0	100
Trihalomethane (mg/L)	raw water	0.2	—
Others			
Cryptosporidium	Very wide variation	0	100
Giardia	depending on the	0	100
Virus	raw water	0	100

UF membranes should be able to remove virus and affect partial removal of trihalomethane potential (THMP) and color. Hybrid processes with UF would give enhanced removal of these species.

2.5 CONCLUSION

Water for domestic use is obtained from surface water sources and/or groundwater sources. Usually, surface water requires a greater degree of treatment compared with groundwater due to the larger amount of impurities present. However, groundwater sometimes requires special water treatment processes, e.g., removal of Fe, Mn, fluoride, etc. Conventional technologies such as filtration, sedimentation, coagulation, etc., have proved to be successful for normal water treatment, while advanced technologies such as activated carbon adsorption, membrane processes, flotation, ion exchange, etc., can be utilized for specific treatment purposes. The suitability of these processes, however, depends primarily on the objective, type of impurity, and mainly on economics. Therefore, laboratory-scale or pilot-scale studies are recommended before the actual application of these technologies.

REFERENCES

IRC, Small Community Water Supplies, International Reference Center for Water Supply and Sanitation, Technical paper series no. 18, Rijswijk, The Netherlands.

Vigneswaran, S. and Ben Aim, R., _Water Wastewater and Sludge Filtration_, CRC Press, Boca Raton, Florida, 1989.

3: Rapid Mixing

CONTENTS

3.1 INTRODUCTION

The purpose of rapid mixing is to disperse coagulant chemicals uniformly throughout the raw water as rapidly as possible in order to destabilize the colloidal particles (i.e., neutralize the negative charges around the colloid surface) present in the raw water. Theoretical and experimental studies have shown that the contact between coagulant and colloidal particles should occur before the hydrolysis reaction with alkalinity is completed. This requires very rapid dispersion of coagulant in the mass of water within a few seconds. To facilitate the rapid dispersion, the water is agitated vigorously with the aid of mixing devices and the coagulant is added at the most turbulent zone. There are basically two ways of effecting rapid mixing:

- hydraulically
- mechanically

The degree of turbulence in hydraulic mixers is a function of flow, i.e., changes in flow will affect mixing efficiencies. Hydraulic rapid mixing is found to be more

41

economical than mechanical rapid mixing due to the absence of moving parts and power requirements, but it is not as flexible as mechanical mixing. Mechanical mixing has been used extensively in both developed and developing countries, but hydraulic mixing is currently receiving increased attention due to recent research and field experiences.

The principles, design, and application of each mixing device is described briefly in this chapter with their relative merits. In general, it is not the velocity gradient (G) that governs the rapid mixing efficiency but the intensity and duration of the agitation (Vrale and Jordan, 1971). Therefore, care must be taken in all rapid mixers to disperse the coagulant as fast as possible.

3.2 HYDRAULIC MIXERS

3.2.1 Principle

The hydraulic mixer is the most basic type of rapid mixer that utilizes the potential head of water for generation of turbulence and eddies, which causes mixing. Most commonly, a hydraulic jump is used for this purpose.

Hydraulic jump is created when flow in an open channel is transferred from supercritical conditions to subcritical conditions, abruptly, the change being accompanied by considerable turbulence and energy loss. A conceptual sketch of hydraulic jump is shown in Figure 3.1. At the junction between the oncoming flow stream and return flow or roller of the hydraulic jump there is a pronounced velocity gradient, and the resulting shear gives rise to the rapid generation of turbulence and the mixing process diffuses all characteristics of flow, both laterally and longitudinally (Rouse et al., 1958). Levy and Ellms (1927) used hydraulic jump for mixing alum with raw water in water treatment processes. They reported that hydraulic jump is an extremely effective means for mixing chemicals with water to be treated.

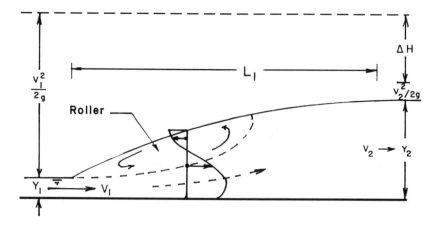

FIGURE 3.1   Conceptual sketch of hydraulic jump. (Rouse et al., 1958)

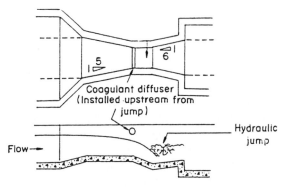

FIGURE 3.2 Typical Parshall flume. (Schulz and Okun, 1984) (Reprinted by permission of John Wiley & Sons, Inc.)

A stable hydraulic jump can be generated by allowing the raw water to flow in a sloping chute followed by a control basin. The supercritical and subcritical flow conditions created in chute and control basin, respectively, induce the jump.

Hydraulic jump mixers are suitable for raw waters that are being coagulated by an "adsorption-bridging" mechanism, because the mixing process in a hydraulic jump involves short mixing time (of the order of 1–2 seconds) and low G values (of the order of 1000 s⁻¹). As considerable headloss (of the order of 0.5 m) takes place in the hydraulic jump mixer, when sufficient natural ground slopes are not available its use becomes uneconomical, because all the treatment units located downstream of the hydraulic jump will have to be placed deeper in the soil. This will result in higher excavation costs, increased soil pressures on the structures, and increased head of water to be lifted.

Parshall flumes, which also use the hydraulic jump turbulence, can be used to achieve rapid mixing. Only the geometry of the device that produces hydraulic jump is different. A detailed sketch of a typical Parshall flume is given in Figure 3.2. It consists of a converging section followed by a diverging section in the open channel. Between these two sections a narrow throat is provided.

3.2.2 Design

Design Parameter Values

Amirtharajah (1978) reports typical residence time (t) and G value, which a hydraulic jump mixer can furnish.

$$t = 2 \text{ s}$$
$$G = 800 \text{ s}^{-1}$$
$$\text{Headloss} = 0.3\text{–}0.4 \text{ m}$$

Design Formulas

A hydraulic jump will form in a channel if the following equation is satisfied (Schulz and Okun, 1984):

$$\frac{Y_2}{Y_1} = \frac{1}{2}\left(\sqrt{1 + 8F^2} - 1\right) \tag{3.1}$$

where Y_1 = upstream depth, Y_2 = downstream depth, F = Froude Number = $v_1/\sqrt{(gY_1)}$, v_1 = upstream velocity.

The location of hydraulic jump on the chute depends upon the depth of tail water and the locus of hydraulic jump. The locus of hydraulic jump is the curve that joins the various points designating the height to which the water would jump at the respective sections. The point where the tail water surface curve intersects the locus of jump is the location of hydraulic jump. For determining the locus of hydraulic jump it is necessary to determine the surface curve on the chute. This curve can be obtained by application of energy equation. Figure 3.3 applies the energy equation between section n and n–1.

$$\left(z_{n-1} + d_{n-1} + \frac{v_{n-1}^2}{2g}\right) - \left(z_n + d_n + \frac{v_n^2}{2g}\right) = \frac{v_{an}^2 N^2}{R_{an}^{\frac{4}{3}}} L_{an} \tag{3.2}$$

where z_{n-1} = bed elevation at section n–1, d_{n-1} = depth of water at section n–1, v_{n-1} = velocity at section n–1, z_n = bed elevation at section n, d_n = depth of water at section n, v_n = velocity at section n, L_{an} = distance between the sections, v_{an} = average of velocities at two sections, R_{an} = average of hydraulic radius at two sections, N = Manning constant. In Figure 3.3, H_n is the head loss in the hydraulic jump.

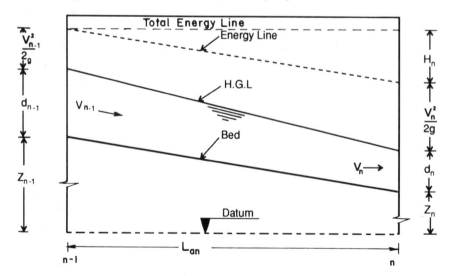

FIGURE 3.3   Definition sketch of nonuniform flow in open channel.

For computation of surface curve, by equation (3.2), a known starting point is required. The depth of water at the beginning of slope of expanding flume is computed, from the formula for broad-crested weir derived from the theory of "maximum flow" (Henderson, 1966).

$$Q = \frac{2}{3}\sqrt{\frac{2}{3g}} \cdot (H)^{\frac{3}{2}} \cdot B \tag{3.3}$$

In metric units,

$$Q = 1.7 \times B\,H^{\frac{3}{2}} \tag{3.4}$$

where Q = flow (m³/s), B = width of flume at entrance (m), H = total energy head upstream of weir (m).

3.2.3 Other Hydraulic Mixers

Weir mixers and turbulent pipeflow-plugflow mixers can also be used as rapid mixers. They are discussed in detail in the literature (Amirtharajah, 1978; Chao and Stone, 1979; Kawamura, 1976; Schulz and Okun, 1984; and Vrale and Jordan, 1971).

The weir used in weir mixers can be of any shape such as V-notch, rectangle, or trapezoid. Flow over the weir causes turbulence and therefore the coagulant is fed where the flow falls from the weir, as shown in Figure 3.4. The baffles that follow the weir help the flow in the subsequent channel to be kept within a tranquil region. Provision of another submerged weir instead of baffles would help a hydraulic jump. The later type is discussed in detail by Schulz and Okun (1984) and it is being used in a water treatment plant in Peru.

Weirs are relatively inexpensive and simple to install. They are also found to be less costly than Parshall flumes. But siltation in the upstream of the weir makes periodic cleaning necessary. Rectangular weirs are more suitable for larger flows and triangular weirs (V-notch) are more suitable for smaller flows. Rectangular weirs are preferred for coagulant mixing because of their better uniform flow distribution over that of triangular weirs.

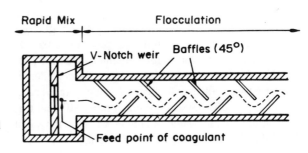

FIGURE 3.4 Baffled channel for rapid mixing. (IRC, 1981)

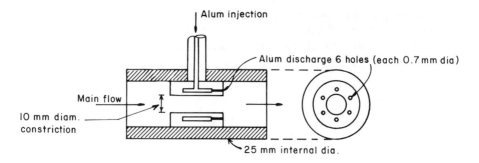

FIGURE 3.5   Typical turbulent pipeflow-plugflow mixer. (Amirtharajah, 1978)

Weir mixers are extensively used in many developing countries, such as India, Brazil, and Kenya. A weir mixer in Nairobi, Kenya makes use of weirs placed along a perimeter of a pedestal, and water flows out radially. The plant treats 250,000 m^3/d of water, which flows through a pressure conduit discharging onto the pedestal. The retention time is about 2–5 seconds. Water flows 1 m over the sharp-crested weir (Schulz and Okun, 1984).

The turbulent pipeflow-plugflow mixer is the simplest of all hydraulic mixers. It involves flow through a pipe with a constricting device to cause turbulence. There are several ways of effecting turbulence; the most convenient method is to incorporate an orifice. Instead of an orifice, other devices such as grids, tapers, baffles, and eductors (Venturi type) can also be used. The turbulent pipeflow-plugflow mixer is so called because of the turbulence caused by a constriction (e.g., an orifice), followed by plug flow in the pipe. It has been found to provide both rapid and slow mixing (Vrale and Jordan, 1971). Figure 3.5 shows a typical turbulent pipeflow-plugflow mixer. If the constriction is an orifice, then the coagulant is dosed just before the orifice. Coagulant is fed through holes in the constriction and diffuses along the flow into the most turbulent region, which is just after the constriction. Flow becomes tranquil further down and the plugflow provides subsequent slow mixing for flocculation. The simplicity of the device is evident.

Chao and Stone (1979) reported on the increased attention given to turbulent pipe mixing. A number of treatment plants utilizing turbulent pipe mixing have been set up (Kawamura, 1976) where adequate mixing is attained within a second. The popularity is mainly due to the practicality, simplicity, and low cost of the turbulent pipeflow mixer. However, turbulent pipeflow mixers have two main disadvantages, viz, limited access for routine maintenance and potential for clogging. Possible remedies for the above drawbacks should be considered while designing such mixers. On the basis of their experimental study, Vrale and Jordan (1971) concluded that turbulent pipeflow mixers are much more efficient than mechanical back-mix mixers, particularly in water treatment where removal of colloids by an adsorption-destabilization mechanism is predominant.

3.3 MECHANICAL MIXERS

3.3.1 Mechanical Back-Mix-Type Reactors

The most commonly used unit for rapid mixing in water treatment plants is the mechanical mixer. Essentially it consists of a tank or vessel made of steel, masonry, or concrete and usually cylindrical or cubical in shape. The vessel is provided with one or more impellers, mounted on an overhung shaft. The shaft is driven by a motor, sometimes directly connected to the shaft but more often connected through a speed-reducing gear box. The impeller creates a flow pattern in the vessel, which eventually results in mixing. The two main types of impeller used for rapid mixing are propellers and turbines.

A propeller is an axial-flow, high-speed impeller whose blades vigorously cut or shear the liquid. The flow currents leaving the impeller cause mixing of coagulant and raw water. A turbine impeller resembles a multibladed paddle agitator with short blades, connected to a shaft mounted centrally in the vessel. The blades may be straight, curved, pitched, or vertical. The impeller may be open, closed, or shrouded. The turbines, while rotating, generate strong currents, radially outward, that destroy stagnant pockets in the vessel.

Generally in rapid mixers, a standard three-blade propeller, open straight-blade or bladed-disk turbine with four to six blades are used.

Application

The mechanical mixers usually provide mixing time in the range of 20 to 60 seconds with G values ranging from 700 to 1000 s^{-1}. The values of G and mixing time provided by mechanical mixers are compatible with the requirements of sweep flocculation, where a prolonged time of mixing furnishes optimum conditions for destabilization of solids through formation of aluminum hydroxide precipitate (Rabbani, 1983).

The mechanical mixers are inefficient and uneconomical in conditions where the predominant mode of destabilization is adsorptive neutralization of colloids by means of hydrolyzed alum ions. Also back-mix conditions in mechanical mixers may result in overdosing of some elements of water while other elements remain underdosed or totally untreated.

The volume of flocculator (V) and power input (P) can be calculated from the following equations.

$$V = Qt \tag{3.5}$$

$$P = \mu V G^2 \tag{3.6}$$

where Q = flow rate (m^3/s), μ = viscosity of fluid (Ns/m^2), G = velocity gradient (s^{-1}), t = rapid mixing time (s).

Table 3.1   Values of K for Equation 3.7 (Rushton, 1952)

Impeller with four baffles at tank wall (each 10% tank diameter)	
Propeller, square pitch, three blades	0.32
Propeller, pitch of two, three blades	1.00
Turbine, six flat blades	6.30
Turbine, six curved blades	4.80
Turbine, six arrowhead blades	4.00
Fan turbine, six blades	1.65
Flat paddle, two blades	1.70
Shrouded turbine, six curved blades	1.08
Shrouded turbine, with stator (no baffles)	1.12

The power requirement of an impeller (rotating in a fluid body under turbulent conditions) can be calculated from the following formula (Rushton, 1952).

$$P = \frac{K}{g} \cdot \gamma \cdot N^3 \cdot D_a^5 \tag{3.7}$$

where P = power requirement (N·m/s), g = specific weight (N/m³), g = acceleration due to gravity (m/s²), N = rotation speed of impeller (rev·/s), D_a = diameter of impeller (m), K = dimensionless constant.

The values of K depend upon the conditions of flow (laminar or turbulent) and type of impeller and are given in Table 3.1. The values of design parameters generally adopted in design of mechanical mixers are as given below:

Mixing time = 20 to 60 seconds.
G values are related to the mixing time inversely. G values relative to the values
    of mixing time, as suggested by ASCE, are given in Table 3.2.
Diameter of mixing tank (D) = 1 to 3 m.
Diameter of impeller (D_a) = 0.2 to 0.4 D.
Speed of flat blade turbine = 10–150 rpm.
Speed of propeller = 150–1500 rpm.

3.3.2 In-Line Blenders

In-line blenders (Figure 3.6) are effectively being used in many industries for rapid mixing. In contrast with the back-mix-type reactors, they are plugflow-type reactors that make use of high-power devices for mixing. They are in the form of standard units,

Table 3.2   Contact Time and Velocity Gradient for Rapid Mixing

Rapid mixing time, s	20	30	40	>40
Velocity gradient, G, s⁻¹	1000	900	790	700

Source: ASCE, AWWA and CSSE, 1969. Reprinted by permission of AWWA.

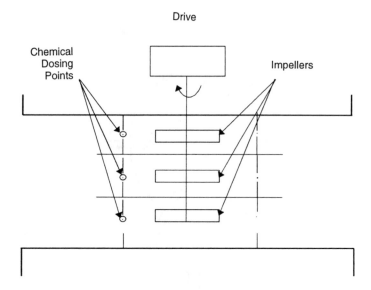

FIGURE 3.6 Schematic representation of in-line blenders.

units, manufactured commercially (fitted in the raw water pipe itself). They can be used in water treatment plants as rapid mixers. Their distinguishing features include very short detention time (on the order of a fraction of a second) and very high G values (3000 to 5000 s^{-1}).

In water treatment processes, the rapid rate of adsorption-destabilization reactions demand instantaneous mixing. This can be attained by using in-line blenders, so their application is most suited when the predominant mode of destabilization is adsorptive in nature, which actually exists when low doses of alum are introduced to high-turbidity water. Hudson and Wolfner (1967) also suggest their use under conditions of adsorption-destabilization.

Some work has been carried out on establishment of design criteria of in-line blenders for use in water treatment processes. Design parameter values suggested by Hudson (1981) are

Power input   = 0.37 kW per 43.8 L/s
Residence time = 0.5 s
Headloss   = 0.3 to 0.9 m
G value   = 3000 to 5000 s^{-1}

3.3.3 Flash Mixer/Injection-Type Mixer

The concept of the injection type mixer is to provide rapid mixing so as to disperse the coagulants in the flowpipe itself. Therefore, turbulent pipeflow-plugflow mixers discussed in Section 3.2.3 could also be classified as belonging to this category. But

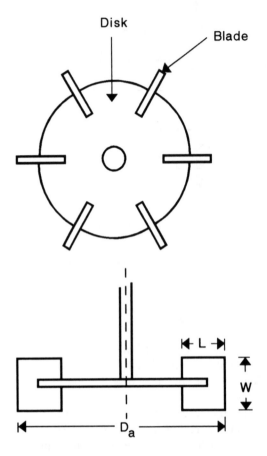

FIGURE 3.7   Bladed disk turbine.

flash mixing, an injection technique, is different in many aspects and deserves to be discussed separately.

The basic principle of flash mixing is to disperse the coagulant irreversibly into the water in a fraction of a second. This is essential when alum is used as the coagulant (Kawamura, 1976). The coagulant is fed into incoming flow through a nozzle by a vertical turbine pump. This causes complete and irreversible dispersion of coagulant into the water by the high speed ejection from the nozzle. Raw water comes in through an influent pipe under pressure (gravitational or pumped) and creates a turbulence just after coagulant injection. This turbulence is sufficient to make the dispersion irreversible. The baffle plate cuts down the turbulence to a sufficient extent to effect subsequent flocculation in the flocculation basin. (See Figure 3.9.) The turbine pump draws water from the raw water pipe through a small branch pipe, mixes it with the alum from the alum storage tank, and ejects the diluted alum solution through the nozzle. It is possible in small treatment plants to eliminate the turbine pump and use the pressure from the main raw water pipe to feed the alum solution. But the alum dosage in such cases must be carefully controlled.

The power input can be calculated using the following formula (Kawamura, 1976):

$$P = \gamma\, q_1(\Delta H) \tag{3.8}$$

where P = power input (Nm/s), γ = specific weight of fluid (N/m^3), ΔH = jet energy loss (m), q_1 = flow rate through nozzle (m^3/s).

since

$$q_1 = C_d\, a\, u \tag{3.9}$$

where C_d = coefficient of discharge = 0.75, a = area of orifice (m^2), u = jet velocity (m/s).

Also

$$\Delta H = \frac{u^2}{2g} \tag{3.10}$$

Substituting equations (3.9) and (3.10) in equation (3.8),

$$P = \frac{\gamma}{2g} \cdot C_d \cdot a \cdot u^3 \tag{3.11}$$

Substituting γ/g = 1000 kg/m^3

$$P = 500 \cdot C_d \cdot a \cdot u^3 \tag{3.12}$$

The following recommended design values can also be used.

G = 750 to 1000 s^{-1}
u = 6 to 7.6 m/s
t = 1 s

Flash mixing is used in several water treatment plants all over the world (Kawamura, 1976). This mixer is suitable for treatment plants of all sizes. Because of its simple operation, it can be used without any problem in developing countries. Because of the high percentage of coagulant utilization, i.e., negligible loss in effective dosage, the coagulant requirement for the same water is less than that for conventional mixers.

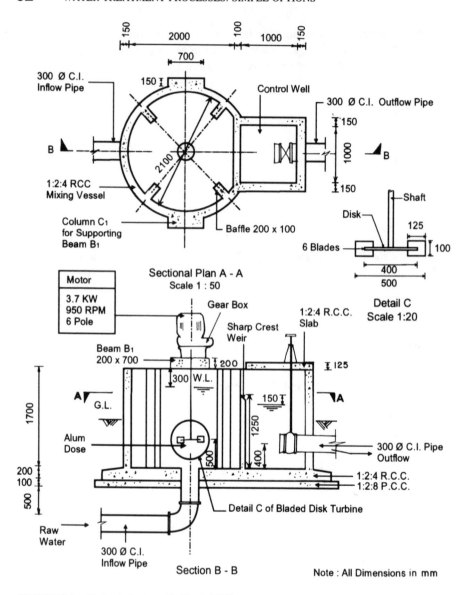

FIGURE 3.8   Mechanical mixer. (Rabbani, 1983)

Injection-type mixers are becoming popular in developed countries because of their better hydraulic performance (rapid and uniform dispersion of coagulant chemicals) in comparison with conventional mechanical back mixers.

3.4 RELATIVE MERITS OF RAPID MIXING TECHNIQUES

The relative merits of the hydraulic and mechanical mixing devices are presented in the Table 3.3.

Table 3.3 Relative Merits of Different Mixing Devices (Adapted from Vigneswaran et al., 1987)

Technique	Advantages	Disadvantages
Hydraulic jump mixers	1. No external power required. 2. Can be constructed with locally available materials. 3. Easily incorporated in the flow channel/pipe itself. 4. Very small detention time. 5. Low maintenance because of the absence of moving parts. 6. Very low head loss. 7. Easy access for maintenance. 8. Easy to construct.	1. Not adjustable for large variations in flow. 2. Considerably affected by upstream flow conditions. 3. Not so effective with large flows (greater than 50,000 m³/d). 4. Cause bed-scouring at the location of jump. 5. Once installed the velocity gradient and mixing time may not be controlled.
Parshall flume mixers	1. Same as those for hydraulic jump (1–7). 2. Can be used as a flow measurement device. 3. Head loss is almost 1/4 of that for weir mixer. 4. Can be used with large flows (e.g., 210,000 m³/d).	1. Same as 1, 2, 4, and 5 for hydraulic jump. 2. Construction not as easy as hydraulic jump mixers. 3. Submergence of the jump may curtail effective power dissipation to mixing.
Weir mixers	1. Same as 1–5 and 7–8 for hydraulic jump mixers. 2. Construction is even simpler than hydraulic jump mixers. 3. The mixing may be controlled to a certain extent by the use of different shapes of weirs. 4. Can be used with large flows (e.g., 250,000 m³/d). 5. Can be used for flow measurement too.	1. Head loss is higher than for hydraulic jump mixers. 2. Silting behind weir necessitates periodic cleaning. 3. Same as 2, 4, and 5 for hydraulic jump mixers.
Turbulent pipeflow-plugflow mixers	1. Simple design. 2. Same as 1–6 for hydraulic jump mixers. 3. Plug flow also provides subsequent flocculation in most cases, thus eliminating need for a separate flocculation basin. 4. Velocity gradient and detention time may be varied by varying turbulence. This is done by changing the orifice (opening) size.	1. Same as 1–3 for hydraulic jump mixers. 2. Clogging of opening is possible. 3. Access for maintenance is not as easy as for the previous cases. 4. Suitable for pipe flow with considerable head (pressure) only. 5. Mixing of coagulant is not visible to operator.
Mechanical backmix	1. Unaffected by flow variations. 2. Lower velocity gradient. 3. Low headloss.	1. More short circuiting. 2. Needs external power. 3. High capital cost.

Table 3.3   (Continued)

Technique	Advantages	Disadvantages
In-line blenders	1. No short circuiting because of plug flow conditions. 2. Reduced cost.	1. Suitable only for high turbid waters.
Flash mixer (injection type)	1. Can be constructed out of locally available materials. 2. Proper mixing does not depend much on G values, therefore accurate control of G values is not necessary. 3. Quite efficient use of coagulant, thus effective savings in chemical cost. 4. Possible to generate high-powered mixing without much energy input. 5. Better control of coagulant dosage. 6. Low detention time. 7. Control is easier with the use of valves.	1. May need external power in large plants. 2. Head losses are higher than for hydraulic jump mixers. 3. Nozzle arrangement is susceptible to clogging and needs periodic cleaning. 4. Requires more maintenance. 5. Capital cost is higher.

3.5 CONCLUSION

It can be noted from the discussion so far that numerous techniques exist for rapid mixing that can be considered low-cost/simple options. Each of the techniques presented has its specific advantages and limitations (Table 3.3). But all of them are certainly applicable in treatment plants with capacities up to 50,000 m³/d. They also incorporate substantially simple to moderately advanced methodologies that are suitable for application in most cases. Past experiences have proved most of them to be cost-effective.

Since most of the techniques do not include accurate mathematical formulation, it is always best to carry out simple lab-scale tests first with the specific water to be treated or to utilize data from similar experiences as guidelines while applying the techniques (Vigneswaran et al., 1987).

EXAMPLE 3.1

Turbidity conditions and jar tests of a raw water show that predominant mode of destabilization is "sweep coagulation," so a mechanical mixer will be the best suited for mixing alum with this raw water. Design a mechanical back-mix-type rapid mixing unit with the following data:

Flow rate = 0.2 m³/s
Temperature (lowest) = 10°C
μ (at 10°C) = 1.3097 × 10⁻³ N·s/m²
γ = 9810 N/m³
g = 9.81 m/s²

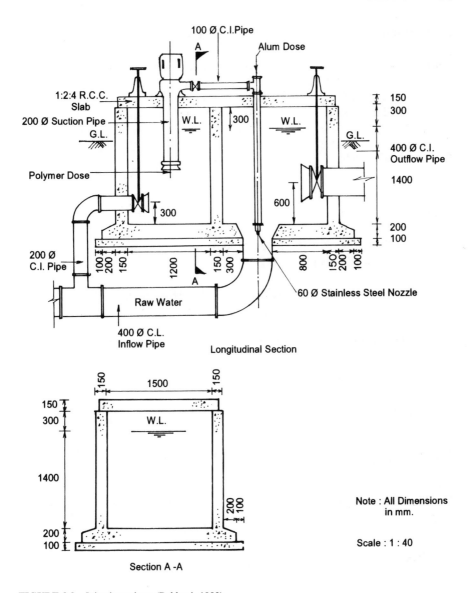

FIGURE 3.9 Injection mixer. (Rabbani, 1983)

Mixing Chamber Design

Provide two chambers of equal size; calculations are for one chamber.

Let t = 45 s.
So V = Qt = 0.1 × 45 = 4.5 m³.

Using 2.1 m diameter, cylindrical tank,

Depth of water = d = $V/(pD^2/4) = 4.5/(p(2.1)^2/4) = 1.3$ m

Let depth d be 1.4 m, to take care of reduction in volume due to overflow weir.
        Free board = 0.3 m
        So depth of vessel = 1.4 + 0.3 = 1.7 m
        Width of baffle = $0.1 \times D = 0.1 \times 2.1 = 0.2$ m

Power Requirement

$$P = \mu \cdot V \cdot G^2$$

From Table 3.2, for t > 40 s, G = 700 sec^{-1}. Therefore,

Power (P) = $1.3097 \times 10^{-3} \times 4.5 \times (700)^2$
          = 2888 N·m/s or W

Power of motor with 80% efficiency = $2888/(0.8 \times 1000)$
                                   = 3.60 kW

Use standard motor of P = 3.7 kW, rpm = 950 and efficiency = 80%.

Impeller Design

From equation (3.7),

$$N = \left(\frac{P \times g}{K \times \gamma \times D_a^5} \right)^{\frac{1}{3}}$$

Use bladed disk turbine with 6 blades. From Table 3.1, value of K_2 for specified turbine is 6.30. Let diameter of impeller = D_a = 0.25 D = $0.25 \times 2.1 = 0.5$ m.

$$N = \left(\frac{2888 \times 9.81}{6.3 \times 9810 \times (0.5)^5} \right)^{\frac{1}{3}} = 2.45 \text{ rev/s} = 150 \text{ rpm}$$

Use gear box to convert 950 rpm to 150 rpm. Other dimensions of impeller are as follows.

E = D_a = 0.5 m
L = $D_a/4$ = 0.5/4 = 0.125 m
W = $D_a/5$ = 0.5/5 = 0.1 m

Design of Overflow Weir

For sharp-crested weir, with $h \gg H$

$$Q = C \cdot B \cdot H^{\frac{3}{2}}$$

where B = breadth of weir (= 1 m), C = weir constant (= 1.80).

$$H = \left(\frac{Q}{C \cdot B}\right)^{\frac{2}{3}} = \left(\frac{0.1}{1.8 \times 1}\right)^{\frac{2}{3}} = 0.145 \text{ m} = 0.15 \text{ m}$$

Design details of this mechanical mixer are shown in Figure 3.8.

EXAMPLE 3.2

For a water, the predominant mode of destabilization is adsorptive and "bridging" by synthetic polymer. Design an injection-type rapid mixing unit (Figure 3.7) with the following data.

Design flow (Q) = 0.2 m³/s
Temperature (min) = 10°C
μ (at 10°C) = 1.3097 × 10⁻³ N·s/m²

Power Input

Using 0.75 kW motor

$$P = \text{Power available at nozzle} = 0.75 \times \eta_p \times \eta_m - h_p$$

where η_p = efficiency of pump = 0.7, η_m = efficiency of motor = 0.7.

Assume h_p = power lost due to pipe friction = 0.027 kW. So P = 0.75 × 0.7 × 0.7 – 0.027 = 0.34 kW = 340 N·m/s.

Nozzle Dimensions

Area of nozzle

$$a = \frac{P}{500 \cdot C_d \cdot u^3}$$

Let u = 7 m/s. Therefore,

$$a = \frac{340}{500 \times 0.75 \times 7^3} = 2.64 \times 10^{-3} \text{ m}^2$$

Diameter of nozzle $[(2.64 \times 10^{-3} \times 4)/\pi]^{0.5} = 0.058$ m = 60 mm.

Injection pump capacity

$$q_1 = C_d \cdot a \cdot u = 0.75 \times 2.64 \times 10^{-3} \times 7 = 14 \text{ L/s}$$

$$G = \sqrt{\frac{P}{\mu V}} = \sqrt{\frac{P}{\mu \cdot Q \cdot t}} = \sqrt{\frac{340}{1.3097 \times 10^{-3} \times 0.2 \times 1}} = 1140 \text{ s}^{-1}$$

The G value is higher than the recommended value of 1000 s^{-1}, but in the adsorptive-destabilization process, a higher value of G improves the mixing efficiency of the mixer. Design details of an injection mixer are shown in Figure 3.9.

EXAMPLE 3.3

Design a rapid mixing unit (in-line blender type) for a design flow of 0.2 m³/s. Turbidity conditions and jar tests of raw water indicate that adsorptive-destabilization takes place during the initial phase of the mixing, so an in-line blender will be the most suitable rapid mixer.

$$\left(\mu = 1.3097 \times 10^{-3} \text{ N} \cdot \text{s/m}^2 \right)$$

Choose a commercial unit of 0.37 kW per 43.8 L/s of flow.

Power input of in-line blender = P = (0.37/43.8) × 200
$$= 1.69 \text{ kW} = 1690 \text{ N·m/s}$$

Volume to which power is supplied = V = 0.5 × 200/10³ = 0.1 m³, where t = 0.5 s

Therefore,

$$G = \sqrt{\frac{P}{\mu V}} = \sqrt{\frac{1690}{1.3097 \times 10^{-3} \times 0.1}} = 3590 \text{ s}^{-1}$$

So the value of G lies within the recommended range of 3000 to 5000 s^{-1}.

REFERENCES

ASCE, AWWA and CSSE, *Water Treatment Plant Design*, American Water Works Association, New York, 1969.

Amirtharajah, A., Design of rapid mix units, in: *Water Treatment Plant Design*, Sanks, R. L. (Ed.), Ann Arbor Science Publishers, Michigan, 1978.

Chao, J. L. and Stone, B. G., Initial mixing by jet injection blending, *J. AWWA*, 71(10), 570–573, 1979.

Chow, V. T., *Open-Channel Hydraulics*, Chapter 15, McGraw-Hill, New York, 1959.

Henderson, F. M., *Open-Channel Flow*, Chapter 6, Macmillan, New York, 1966.

Hudson, H. E., Jr. and Wolfner, J. P., Design of mixing and flocculation basins, *J. AWWA*, 59, 1257, 1967.

Hudson, H. E., Jr., *Water Clarification Process*, Chapter 6, Van Nostrand Reinhold, New York, 1981.

IRC, International Reference Center for Water Supply and Sanitation, *Small Community Water Supplies*, Technical Paper Series 18, Rijswijk, The Netherlands, 1981.

Kawamura, S., Considerations on improving flocculation, *J. AWWA*, 68(6), 328–336, 1976.

Levy, A. G. and Ellms, J. W., The hydraulic jump as a mixing device, *J. AWWA*, 17(1), (1927).

Rabbani, W. I., *Techno-Economic Comparison Among Various Types of Rapid Mixers Used in Water Treatment Plants*, RSPR-EV-83-3, Asian Institute of Technology, Bangkok, 1983.

Rouse, H., Siao, T. T. and Nagaratnam, S., Turbulence characteristics of the hydraulic jump, *ASCE J. Hyd. Div.*, Vol. 84, Paper 1528, 1–30, 1958.

Rushton, J. H., Mixing of liquids in chemical processing, *Ind. Eng. Chem.*, 44, 2931, 1952.

Schulz, C. R. and Okun, D. A., *Surface Water Treatment for Communities in Developing Countries*, John Wiley & Sons, New York, 1984.

Vigneswaran, S., Shanmuganantha, S. and Mamoon, A. A., Trends in water treatment technologies, *Environmental Review*, No. 23/24, ENSIC, Thailand, 1987.

Vrale, L. and Jordan R. M., Rapid mixing in water treatment, *J. AWWA*, 63, 52, 1971.

4: Flocculation

CONTENTS

4.1 INTRODUCTION

Flocculation is a process used in water treatment for aggregation or growth of destabilized colloidal particles, which can be easily removed through subsequent treatment methods such as sedimentation or filtration. The three major mechanisms of flocculation are:

1. Aggregation resulting from random Brownian movement of fluid molecules (perikinetic flocculation). When particles move in water under Brownian motion, they collide with other particles. On contact, they form large particles and continue to do so until they become too large to be affected by Brownian motion. Perikinetic flocculation is predominant for submicron particles. A large initial concentration of particles in the suspension will cause faster floc formation, since the opportunity for collision is higher.

2. Aggregation induced by velocity gradient in the fluid (orthokinetic flocculation). Orthokinetic flocculation that involves particle movement with gentle motion of water considers that particles will agglomerate if they

61

collide and become close enough to be within a zone of influence of one another. It also considers that particles have negligible settling velocity; hence there is a need for agitation of the water, or a velocity gradient to promote the collisions. The rate of flocculation is proportional to the velocity (shear) gradient, the volume of the zone of influence, and the concentration of particles.

3. Differential settling, where flocculation is due to the different rates of settling of particles of different sizes. Larger particles settle faster than smaller particles, which makes the relative velocities between the particles different. This also helps in orthokinetic flocculation because velocity gradients are produced, causing further agglomeration.

The two main modes of process operations used in flocculation (i.e., creation of velocity gradients) are

1. Hydraulic flocculators: hydraulic energy provides the necessary velocity gradient.
2. Mechanical flocculators: the velocity gradient is created by mechanical power input.

Mechanical flocculators use mechanical mixing devices such as paddles, turbines, propellers, etc. Although mechanical flocculators are used widely, they require extensive maintenance. Mechanical flocculators are discussed in detail in standard textbooks dealing with water treatment processes.

Due to the difficulties experienced with these mechanical flocculation techniques, flocculators using hydraulic energy have gained prominence in the developing countries. The most widely used hydraulic flocculator is the baffled channel flocculator (Schulz and Okun, 1984). It is widely used in many developing countries and performs efficiently over a wide range of flows. In this chapter, the baffled channel flocculator is discussed in brief with an example, and detailed discussions can be found in standard text books on water technology. More emphasis is given in this chapter to nonconventional, hydraulic-type flocculators such as gravel bed flocculators and Alabama flocculators.

4.2 BAFFLED CHANNEL FLOCCULATOR

With a baffled channel flocculator, mixing is accomplished by the change in direction of flow of water through channels. Basically, there are two types of baffle systems: vertical baffling type (Figure 4.1) and horizontal baffling type (Figure 4.2). The change in direction of flow creates headloss in the channels, which in turn creates a velocity gradient for slow mixing. This is governed by the following equations.

$$G = (Q \rho g \Delta H / \mu V)^{1/2} \tag{4.1}$$

where g = acceleration due to gravity, Q = flow rate, V = volume of the flocculator, ρ = density of fluid, μ = viscosity of fluid, ΔH = headloss.

CROSS SECTION

FIGURE 4.1   Vertical baffling type.

For the vertical baffling type,

$$\Delta H = h_b + h_c + h_o \tag{4.2}$$

where $h_b = f_b v^2/2g$ = headloss in the lower bend, $h_c = l\, v^2/C^2R$ = friction loss in closed conduit, $h_o = v^2/2g$ = headloss in the overflow, f_b = coefficient of headloss in the bend (general range from 2.5–3.5), l = length of the route travelled, C = Chezy's coefficient = $R^{1/3}/n^2$, n = coefficient of roughness, R = hydraulic radius, v = flow velocity in the basin.

For the vertical baffle type, the R calculation is made assuming the system as an open conduit, whereas for the horizontal baffle type, it is considered as a closed conduit.

The typical design values of the baffled channel flocculator are summarized in Table 4.1. The advantages and limitations are discussed in Table 4.2.

4.3 GRAVEL BED FLOCCULATOR

In view of the complications involved in operating a full-fledged flocculator-clarifier-filter chain individually, in developing countries the gravel bed flocculator seems to be a viable alternative to achieve flocculation and partial solid separation in small community water supplies.

4.3.1 Basic Concepts

It is well known that flocculation is effected by mixing energy (provided through the velocity gradient). In conventional flocculators, a mechanical device or a hydraulic device provides the velocity gradient. In a gravel bed flocculator, the necessary velocity gradient is provided by the sinuous flow of water through the gravel bed

PLAN

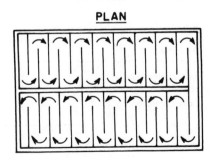

FIGURE 4.2   Horizontal baffling type.

Table 4.1   Design Criteria for the Baffled Channel Flocculator

Design parameters	Range of values
Distance between baffles	Not less than 0.45 m
Clear distance between baffles and walls	Generally 1.5 times of distance between baffles
Depth of water	Not less than 1 m
Velocity of fluid	0.15–0.45 m/s
**Suitable arrangement should be made within the tank to remove the scum and sludge produced.	

either in the upward or downward direction. The velocity gradient produced is a function of the size of the gravel, rate of flow, cross-sectional area of the bed, and the head loss across the bed. Gravel bed flocculators should always be followed by further treatment (e.g., sedimentation and/or filtration) because their purpose is to produce a suitable size of floc for removal by subsequent treatment units. Any floc removal in the flocculator itself is incidental to its main purpose. However, for high turbid wastes, the solid removal efficiency has been noticed to be significantly high (Thanh, 1985).

4.3.2 Process Description

A schematic diagram of the treatment train using a gravel bed flocculator is given in Figure 4.3.

The raw water with the predetermined alum dosage (determined through a jar test) enters the gravel bed flocculator through the bottom or top, depending on whether it is an upflow or downflow type. A typical gravel bed flocculator is shown in Figure 4.4. The bed may consist of either uniform or varying sizes of gravel. A uniformly graded gravel bed as shown in Figure 4.4 is advantageous, as it functions as a tapered flocculator to give initial rapid mixing and subsequent slow mixing to form big flocs. (The tapered flocculator is a popular type of mechanical mixing flocculator. Here the flocculator tank is divided into several chambers, usually three to four chambers. The mixing intensity is high in the first chamber, and then it drops gradually in the subsequent chambers. The high mixing speed in the first chamber creates a condition to form quick flocs. The lower mixing speed in the subsequent chambers provides favorable conditions for the formation of larger flocs without floc breakage. Tapered flocculation with mechanical mixing is discussed in brief with some typical design data and with an example at the end of this chapter.)

Table 4.2   Advantages and Limitations of the Baffled Channel Flocculator

Advantages	Disadvantages and limitations
Less energy cost	High headloss
No mechanical or moving parts	Lack of flexibility of mixing intensity
Less maintenance	Performs well only if plant flow rate and water quality are constant

FIGURE 4.3 Schematic diagram of treatment train.

In a downflow-type clarifier as shown in Figure 4.4, the size of gravel is uniformly increased along the flow, i.e., in the downward direction, and the velocity gradient decreases with the increase in gravel size (equation (4.3)). Therefore, it should be realized from Figure 4.4 that a higher velocity gradient, i.e., rapid mixing, is given initially and a low velocity gradient, i.e., slow mixing, is given subsequently. The gravel bed flocculator may be followed by a clarifier and filter if the effluent turbidity and color from the flocculator is beyond the acceptable limit of effluent quality. The sedimentation tank is generally omitted in smaller treatment plants and the effluent from the gravel bed flocculator is sent directly to the filter (Schulz and Okun, 1984) as in the case of direct filtration.

The gravel bed pores get filled up with flocs leading to a gradual increase in head loss, which is generally higher than in the other flocculators. Periodic cleaning of the gravel bed similar to backwashing of the filters is therefore necessary. The backwashing is in a fixed bed state but with a high rate of backwashing. The frequency of cleaning will vary from plant to plant, depending on the raw water characteristics.

4.3.3 Design Criteria

As stated in Section 4.3.1, the velocity gradient that governs the flocculation depends on the

1. size of the gravel
2. rate of flow
3. cross-sectional area of the bed
4. head loss across the bed

The above dependencies can be stated through the following relationships, as obtained from Fair et al. (1971) and Schulz and Okun (1984) assuming laminar flow conditions across gravel bed:

$$G = [\Delta H \rho g Q / \mu \varepsilon V]^{1/2} \qquad (4.3)$$

and the head loss (ΔH) is given by the Carmen-Kozeny equation,

$$\Delta H = (f/\Theta)([1 - \varepsilon]/\varepsilon^3)(L/d)(v^2/g) \qquad (4.4)$$

$$f = 150\{(1 - \varepsilon)/Re\} + 1.75 \qquad (4.5)$$

$$Re = d v \rho / \mu \qquad (4.6)$$

where G = velocity gradient, s^{-1}; DH = head loss, m; r = specific gravity of water, kg/m^3; g = acceleration due to gravity, m/s^2; Q = flow rate, m^3/s; m = dynamic viscosity, kg/m◊s; e = porosity (. 0.4); V = volume of gravel bed, m^3; f = friction factor; L = depth of the gravel bed, m; Q = shape factor (approximately equal to 0.8); Re = Reynolds number; d = average size of gravel, m; v = approach velocity, m/s.

Pilot-scale studies may be conducted, relatively inexpensively, if greater accuracy is needed. But for small treatment plants, the design using the above equations has proved to be adequate (Schulz and Okun, 1984). Full-scale plants have produced results comparable to those of jar tests and pilot-scale tests at the Iguacu plant in Curitiba, Brazil (Schulz and Okun, 1984).

Tapered upward-flow gravel bed flocculators are also used to take more advantage of the tapered flocculation. The concept is the same as that for the downward-flow type. But the tapered flocculator shown in Figure 4.5 is so designed that the velocity gradient decreases along the flow. The gravel size need not be varied beyond two sizes (Bhole, 1981).

With both upward-flow and downward-flow types, a sludge drain to drain the sludge during backwashing and a hopper bottom to collect the sludge for easy removal are provided (Figures 4.4 and 4.5). The gravel bed is generally supported on a corrosion-free wire mesh or grid.

4.3.4 Application Status

There seems to be no evidence of extensive research on gravel bed flocculators (Thanh, 1985), but they are being used extensively in developing countries, particularly in India (Badrinath, 1977; Schulz and Okun, 1984) and Latin America (CEPIS, 1982). Gravel bed flocculators were successful in India mainly due to the simplicity, low cost, and the effective method of flocculation. Gravel bed flocculators were also found to be successful as low-cost package plants in Brazil. This flocculation process can be utilized in remote areas where semi-skilled labor and the necessary materials are scarce and the community is quite small. Package plants are also well suited for small communities or in remote areas where on-site construction is difficult.

Some typical values for gravel bed flocculator parameters are given in Table 4.3. The values given in Table 4.3 are either from full-scale operations or laboratory studies, mostly in India, Latin America, and Thailand.

Package plants (which consist of gravel bed flocculation, tube settling, and dual media filtration) with a capacity of 270 m^3/d were found to reduce turbidities of the order of 500 NTU to 10 NTU in India (Bhole, 1977). The size of the above plant is $5.3 \times 1.25 \times 1.25$ m^3 and it weighs 1.3 tons. Assuming a per capita consumption of 40 L/d (only for drinking purposes), the above package plant is sufficient for 2000 people. This plant was found to cost about U.S. $2500 (1982 value). Another application was found in treating Red Hill Lake waters in Madras, India, to remove color and turbidity (20–30 units) (Badrinath, 1977).

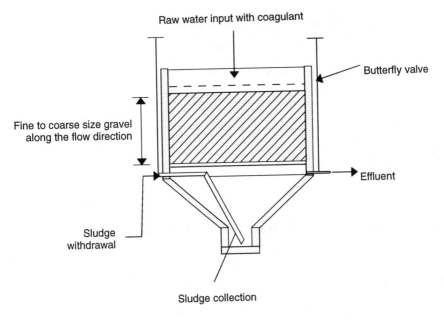

FIGURE 4.4 Downward-flow gravel bed flocculator. (Kardele, 1981)

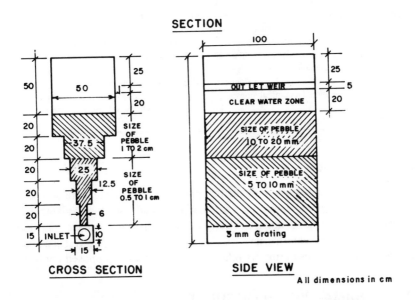

FIGURE 4.5 Upward-flow gravel bed flocculator. (Modified from Bhole, 1981)

Table 4.3   Typical Gravel Bed Flocculator Parameters

Parameter	Unit	Value	Reference
Flow rate	m³/h	10–200	
Velocity gradient			
Rapid	s⁻¹	130–1230	Schulz and Okun (1984)
Slow	s⁻¹	35–40	Vigneswaran et al. (1987)
Alum dose	mg/L	20–60	Vigneswaran et al. (1987)
Flocculation time	min	3–5	Schulz and Okun (1984)
Gravel size	mm	10–60	Schulz and Okun (1984)
Raw water turbidity	NTU	10–300	Schulz and Okun (1984)
Bed depth	m	1.5–3.0	Schulz and Okun (1984)
Head loss	m	0.01–0.2	Schulz and Okun (1984)

4.3.5   Relative Merits of Gravel Bed Flocculators

Table 4.4   Merits and Limitations of Gravel Bed Flocculator

Advantages	Limitations
Simple and inexpensive design for flocculation.	Needs periodic cleaning and therefore plant and maintenance labor are needed.
Experience in developing countries has proved their suitability beyond doubt.	Fouling due to biological growth on gravel bed is possible since sludge deposits on gravel.
Well-suited as package plants, as they are compact.	Not suitable for large plants due to limited capacity. However, it should be noted that the capacity can be increased by using multiple units in parallel.
Low maintenance; no mechanical devices are involved.	
Act as pretreatment units prior to filtration by trapping most of the larger flocs and often help in eliminating clarifiers.	Head loss is larger than for conventional flocculators.
Upflow or downflow modes can be used as appropriate according to site conditions.	Dosage of flocculent-aid, if found necessary, would be a problem since a lag time between dosages of flocculent and flocculent-aid is necessary for efficient flocculation.
Flocculation time is considerably lower than that for conventional flocculation.	
Flocs retained in the gravel bed enhance flocculation, resulting in reduced chemical consumption.	Anaerobic conditions will set in in the lower part, causing odor problems and deterioration of effluent quality.
Construction and operation require minimal skill.	

4.3.6   Conclusion

The above discussions clearly indicate that gravel bed flocculators are best suited for application in developing countries, particularly in smaller treatment plants (less than or equal to 5000 m³/d). Advantages clearly outweigh the disadvantages or, more precisely, the limitations. Again, like any water treatment technique, the gravel bed flocculator must be evaluated, considering its relative merits, for each proposed application.

4.4   ALABAMA-TYPE FLOCCULATORS

Another innovative technique in flocculation is the Alabama-type flocculator. IRC (1981) cites this technique as an appropriate one for developing countries, particularly for small communities. The energy for flocculation is hydraulically transmitted.

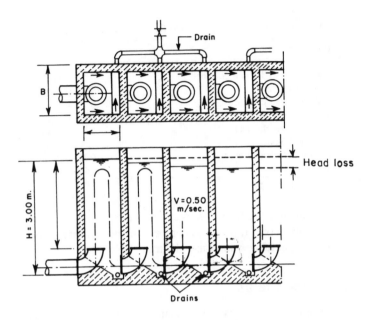

FIGURE 4.6 Alabama-type flocculator. (IRC, 1981)

There are several methods of hydraulic flocculation, but the Alabama-type flocculator is the most suitable for small-scale application. It has, however, yet to be applied extensively in developing countries. It was developed in Alabama, hence its name, and later used in Latin America.

4.4.1 Process Description

There is no specific theory behind the Alabama-type flocculator, although design guidelines are available based on past experience. The process incorporates a special flocculator, which is shown in Figure 4.6

There are several chambers placed in series and the number and size of chambers vary from application to application. Water enters the first chamber under pressure, mechanical or gravitational, through an inverted elbow as shown in Figure 4.6. Flow takes place in two directions in each chamber, i.e., upwards and downwards. Flocculation takes place due to the turbulence caused during these upward and downward flows. Water enters subsequent chambers in a similar manner, one after the other.

The chamber can be built out of brickwork or concrete depending on the capacity and soil type. The pipes can be either plastic or cast iron. Coagulant is fed before the water enters the flocculator.

4.4.2 Design Criteria

Design is based on past experience, and sufficient guidelines for design are available. Some typical values for the chamber dimensions and the flocculation parameters are given in Table 4.5.

Table 4.5 Typical Design Parameters for Alabama-Type Flocculators

Parameter	Value
Rated capacity for unit chamber	25–50 L
Velocity at turns	0.40–0.60 m/s
Length of unit chamber	0.75–1.50 m
Width of unit chamber	0.50–1.25 m
Depth of unit chamber	2.50–3.50 m
Detention time	15–25 min
Head loss for entire unit	0.35–0.50 m
Velocity gradient	40–50 s^{-1}

Source: IRC, 1981.

Care must be taken to produce effective flocculation by providing at least 2.5 m of water above the outlet pipe level. The chamber dimensions vary according to flow rate. Some practical guideline values are given in Table 4.6.

4.4.3 Application Status

The functioning of the Alabama-type flocculator seems to be quite satisfactory; the flocculator is economical to construct as well as to operate and maintain, needing very nominal supervision and operation. Short-circuiting can be a problem with these units. Drains should be provided in each chamber for cleaning, since material tends to accumulate at the bottom. The only maintenance to be done quite frequently (once a week or so) is to remove the sludge that accumulates at the bottom through the drains (see Figure 4.6) provided for this purpose.

4.4.4 Conclusion

The Alabama-type flocculator is suitable for small plants in developing countries mainly due to its simplicity and cost effectiveness. But care must be taken in relying

Table 4.6 Guideline Values for Alabama-Type Flocculator Chamber Dimensions

Flow rate Q (L/s)	Width B (m)	Length L (m)	Diameter D (mm)	Unit chamber area (m²)	Unit chamber volume (m³)
10	0.60	0.60	150	0.35	1.1
20	0.60	0.75	250	0.45	1.3
30	0.70	0.85	300	0.60	1.8
40	0.80	1.00	350	0.80	2.4
50	0.90	1.10	350	1.00	3.0
60	1.00	1.20	400	1.20	3.6
70	1.05	1.35	450	1.40	4.2
80	1.15	1.40	450	1.60	4.8
90	1.20	1.50	500	1.80	5.4
100	1.25	1.60	500	2.00	6.0

Source: IRC, 1981.

on its capability to treat highly turbid and colored waters because the control of floc formation is highly limited. Therefore, subsequent removal of flocs in clarifiers may not be efficient and may lead to the blockage of filter pores. But as long as its use is limited within the guidelines discussed in the preceding sections, the Alabama-type flocculator will function efficiently.

EXAMPLE **4.1**

Design a tapered flocculator with mechanical mixing arrangement for a design flow of 550 L/s, using the following data.

- Number of compartments = 4
- Total flocculation time = 20 min
- Optimum alum dose = 38.3 mg/L
- Mixing device = 3 blade propeller

Solution

The empirical formula relating flocculent dose and flocculation design parameters (Letterman et al., 1973) is

$$G^{2.8} t = 44 \times 10^5/C$$

where t = flocculation time (min), G = velocity gradient (s^{-1}), C = optimum flocculent concentration (mg/L). Assuming optimum alum dose (C) = 38.3 mg/L, therefore,

$$G^{2.8} t = 1.149 \times 10^5$$

or

$$G = 22\,s^{-1}$$

Power input (P) = $k \rho N^3 D^5$

where k = constant = 1 (for three blade propellers), N = number of revolutions of blades/min, ρ = density of water.

Check Camp's condition for successful flocculation,

$$2 \times 10^4 < Gt < 2 \times 10^5$$
$$Gt = 22 \times 20 \times 60$$
$$= 26400, \text{ thus OK}$$

Adopt flocculation tank layout given in Figure 4.7.

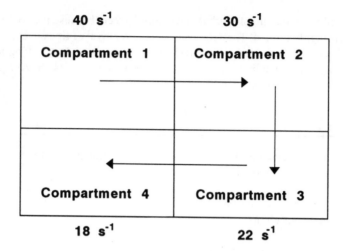

$$40 \ s^{-1} \qquad\qquad 30 \ s^{-1}$$

| Compartment 1 | Compartment 2 |
| Compartment 4 | Compartment 3 |

$$18 \ s^{-1} \qquad\qquad 22 \ s^{-1}$$

FIGURE 4.7 Flocculation tank layout. (Amirtharajah, 1978)

Volume of each tank = flow × flocculation time
$$= 0.55 \times 5 \times 60$$
$$= 165 \ m^3$$
$$= 8 \ m \ length \times 8 \ m \ width \times 3 \ m \ depth$$

Provide variable speed motor that can operate up to $G = 75 \ s^{-1}$. At

$$G = 75 \ s^{-1}$$
$$P = G^2 \mu V$$
$$= 75^2 \times 1.38 \times 10^{-3} \times 8 \times 8 \times 3$$
$$= 1.49 \ kW \approx 1.5 \ kW$$

Assuming 75% efficiency and a flocculator paddle diameter of 20% of tank width (i.e., 8 × 0.2 = 1.6 m), for a three blade propeller (as taken by Amirtharajah, 1978), k = 1

$$P = k \rho N^3 D^5$$
$$0.75 \times 1500 = 1 \times 1000 \times N^3 \times 1.6^5$$
$$N = 28.5 \ rpm$$

Using a variable-speed motor (to obtain a range of G values from 15 to 75 s^{-1}),

$$P = G^2 \mu V$$
$$P = 15^2 \times 1.38 \times 10^{-3} \times 8 \times 8 \times 3$$
$$= 60 \ W$$

Assuming an efficiency of 75%, the rotation speed of the propeller at $G = 15 \text{ s}^{-1}$,

$0.75 \times 60 = 1 \times 1000 \times N^3 \times 1.6^5$
Hence $N = 10$ rpm

Therefore, supply a 1.5 kW variable-speed motor geared to operate between 10 and 28.5 rpm.

Tapered flocculation is popular in small- and medium-sized water treatment plants. Some of the tapered flocculation facilities in New South Wales, Australia, are listed in Table 4.7.

EXAMPLE 4.2

Design a horizontal baffle-type flocculator for a design flow (Q) of 48 million liters per day (MLD) using the following design values.

- Detention time (t) = 30 min
- Flow velocity through channels (v_1) = 0.3 m/s
- Distances between baffles (L_B) and between the end of baffle and the wall (L_w) are 0.45 m and 0.7 m respectively
- Width of flocculation tank (W) = 8 m
- Thickness of each baffle = 0.075 m

Solution

Volume of flocculation (V) = Q × t
$$= (48 \times 10^3 \text{ m}^3 / 24 \times 60 \text{ min}) \times 30 \text{ min} = 1000 \text{ m}^3$$
Total length of flow $(L_F) = v_1 \times t$
$$= 0.3 \times 30 \times 60 = 540 \text{ m}$$
Depth of water in the flocculator $= V/(L_F \cdot L_B)$
$$= 1000/(540 \times 0.45) = 4.2 \text{ m}$$
Effective length of each baffle $= L_e = W - L_w$
$$= 8 - 0.7 = 7.3 \text{ m}$$
Number of channels required = 540 m / 7.3 m = 74
$$= 37 \text{ channels in each half of the tank}$$
Overall length of the tank = The clear length + total wall thickness
$$= 37 \times 0.45 + 36 \times 0.075 = 19.35 \text{ m}$$

EXAMPLE 4.3

Design a gravel bed flocculator for a design flow of 5 L/s (432 m³/d). The configuration and gravel sizes and depths of gravel bed flocculator were given in Figure 4.5. Assume the sectional width to be 160 cm.

Table 4.7 Selected Flocculation Facilities and Design Criteria in New South Wales, Australia

Plant	Capacity (ML/d)	Flocculator type	Number of compartments	G value (s⁻¹)	Power (kW)	Detention time (min)	Coagulant/ coagulant aid/ flocculent aid
Guyra TP	5.5	Tapered flocculation (equipped with vertical shaft mixers)	4 ($12 \times 3.5 \times 3.2\ m^3$)	200/75/25	1.6/0.2/0.02	33 (8.25 min each)	alum/ soda ash/ polyelectrolyte
Lithgow TP	15	Tapered flocculation (equipped with horizontal shaft mixers)	3 ($9.3 \times 9.3 \times 3.3\ m^3$)	200/125/25	3.8/1.48/0.06	25 (8.33 min each)	alum/ soda ash/ polyelectrolyte
Copeton TP	23	Two sets of tapered flocculation (equipped with vertical shaft mixers)	3 ($3.9 \times 3.9 \times 3.5\ m^3$)	80/50/25	0.6	20 (6.67 min each)	alum/ polyelectrolyte
Bamarang TP	60	Two sets of tapered flocculation (equipped with horizontal shaft mixers)	Two sets of 3 compartment tanks	100/50/20	4/1/0.15	30 (10 min each)	alum/ soda ash/ polyelectrolyte
Gosford Wyong TP	146	Stage 1 tapered flocculation	3	80/50/20	2.2/1.1/0.37	33 (11 min each)	alum/ soda ash/ polyelectrolyte
		Stage 2 tapered flocculation (horizontal shaft mixers)	4	78/47/20/11	5.5/3/1.1/0.37	44 (11 min each)	alum/ soda ash/ polyelectrolyte

Source: Adapted from Cowele, 1992.

Additional data:

- Water viscosity (μ) and density (ρ) are 1.01×10^{-3} kg/m·s and 1000 kg/m^3 respectively
- Porosity of gravel bed (ε) is 0.4
- Shape factor of gravel is 0.8

Solution

Volume of flocculator $= 1.6 \times 0.2 \times (0.06 + 0.125 + 0.25 + 0.375 + 0.5)$
$$= 0.42 \text{ m}^3$$
Total flocculation time (t) $= V/Q = 0.42$ m^3/5 L/s $= 84.0$ s
$$= 1.4 \text{ min}$$

Headloss and gradient calculation in section

Approach velocity (v) $= 5$ L/s $/(1.6 \times 0.06)$ m^2
$$= 0.052 \text{ m/s}$$
Reynolds Number (Re) $= d\, v\, \rho\, / \mu$

where d = average diameter = $(0.005 + 0.010)/2 = 0.0075$ m, v = 0.052 m/s. Therefore

Re $= 0.007 \times 0.052 \times 1000\, /\, 1.01 \times 10^{-3} = 386$

The friction factor is given by the following formula

Friction factor (f) $= 150 \times ([1 - \varepsilon]/\text{Re}) + 1.75$
$$= 150 \times (1 - 0.4/386) + 1.75$$
$$= 1.98$$

Headloss through the first layer can be calculated from Kozeny's formula as given below. (The details of the equations were given in Section 4.3.3)

$$\Delta H = (f/\Theta)\left([1 - \varepsilon]/\varepsilon^3\right)(L/d)\left(v^2/g\right) \qquad (4.4)$$
$$\Delta H = (1.98/0.8) \times (1 - 0.4)/0.4^3 \times (0.2/0.0075) \times \left(0.052^2/9.81\right)$$
$$= 0.17 \text{ m}$$

Volume (V) of the first section $= 1.6 \times 0.06 \times 0.2 = 0.0192$ m^3
Velocity gradient (G) $= [\Delta H\, \rho\, g\, Q\, /\, \mu\, \varepsilon\, V]^{1/2}$
$$= \frac{\left(0.17 \times 1000 \times 9.81 \times 5 \times 10^{-3}\right)^{1/2}}{\left(1.01 \times 10^{-3} \times 0.4 \times 0.0192\right)^{1/2}}$$
$$= 1037 \text{ s}^{-1}$$

In Sections 2, 3, 4 and 5, the G values can be calculated using the same procedure. The headloss and velocity gradient values are summarized below.

Section 1: $\Delta H = 0.17$ m; $G = 1037$ s^{-1}
Section 1: $\Delta H = 0.045$ m; $G = 369$ s^{-1}
Section 1: $\Delta H = 0.014$ m; $G = 146$ s^{-1}
Section 1: $\Delta H = 0.0027$ m; $G = 52$ s^{-1}
Section 1: $\Delta H = 0.0017$ m; $G = 36$ s^{-1}

The velocity gradient ranges from 1037 s^{-1} at the inlet where rapid mixing (uniform dispersion of flocculent in the water mass) occurs, to 36 s^{-1} in the uppermost and largest section where slow mixing or flocculation occurs.

EXAMPLE 4.4

Calculate the dimensions of an Alabama-type flocculator for the following data.

Flow, $Q = 20$ L/s
Detention time $= 15$ min
Size of curved pipe $= 250$ mm (10")

From Table 4.6, unit chamber measures $0.60 \times 0.75 \times 2.5$ m^3

Volume of unit chamber $= 1.3$ m^3
Total volume required $= 15 \times 1.2 = 18$ m^3
Number of chambers $= 18/1.3 = 14$

REFERENCES

Amirtharajah, A., Design of flocculation systems, *Water Treatment Plant Design for Practicing Engineers*, R. L. Sanks (Ed.), Ann Arbor Science, Michigan, 1978.

Badrinath, S. D., An economic upflow unit for surface waters with low turbidity, *J. Indian Water Works Association*, 9(3), 233, 1977.

Bhole, A. G., Study of low cost sand-bed-flocculator for rural areas, *Indian J. Environ. Health*, 19(1), 38, 1977.

Bhole, A. G., Design and fabrication of low cost package water treatment plant for rural areas in India, *AQUA*, (5), 315–320, 1981.

CEPIS, *Modular Plants for Water Treatment*, Vol. 1 and 2, Technical Document No. 8, Pan American Health Organization, Lima, Peru, 1982.

Cowele, S., *Cost analysis of water treatment plants in New South Wales*, Final year thesis report, University of Technology, Sydney, Australia, 1992.

Fair, G. M., Geyer, J. C. and Okun, D. A., *Elements of Water Supply and Wastewater Disposal*, John Wiley & Sons, New York, 1971.

IRC-International Reference Center for Water Supply and Sanitation, *Small Community Water Supplies*, Technical Paper Series 18, Rijswijk, The Netherlands, 1981.

Letterman, R. D., Quon, J. E. and Gemell, R. S., Influence of rapid-mix parameters on flocculation, *J. AWWA.*, 65(11), 716–722, 1973.

Schulz, C. R. and Okun, D. A., *Surface Water Treatment for Communities in Developing Countries*, John Wiley & Sons, New York, 1984.

Thanh, T. T., *A detailed study of gravel bed filter performance*, Masters thesis report, Asian Institute of Technology, Bangkok, 1985.

Vigneswaran, S., Shanmuganantha, S. and Mamoon, A. A., Trends in water treatment technologies, *Environmental Sanitation Reviews*, No. 23/24, December 1987, AIT, Bangkok.

5: Sedimentation

CONTENTS

5.1 INTRODUCTION

Sedimentation is a solid-liquid process, making use of the gravitational settling principle. In water treatment plants, sedimentation is used to remove settleable solids. Since the size of the particles in the surface water is smaller, sedimentation is preceded by flocculation.

The efficiency of the sedimentation process is related to various factors: loading rate, water quality, particle/floc size and weight, tank geometry, etc. A sedimentation tank can be designed for optimum efficiency (90–95% floc removal) or can be designed to operate at lower efficiencies, allowing the filters to remove more of the solids. Usually the latter approach leads to a total plant optimization.

Particles settle from suspension in different ways, depending on the characteristics and concentration of the particles. Four distinct types of sedimentation have been classified, reflecting the influence of the concentration of the suspension and the flocculating properties of the particles:

1. Class-1 clarification: Settling of dilute suspensions that have little or no tendency to flocculate.
2. Class-2 clarification: Settling of dilute suspensions with flocculation taking place during the settling process.
3. Zone settling: Particles settle as a mass and not as discrete particles. Interparticle forces hold the particles (which are sufficiently close) in a fixed position, so that the settlement takes place in a zone.
4. Compression settling: Settlement takes place over the resistance provided by the compacting mass resulting from particles that are in contact with each other.

5.2 CONVENTIONAL SEDIMENTATION TANK

In a conventional sedimentation tank, the flow is usually horizontal. Circular or rectangular configurations are common for sedimentation tanks with horizontal flow (Figures 5.1a and b).

The main design criteria for a sedimentation tank with horizontal flow are the surface loading rate (Q/A), adequate depth and detention time for settling, and suitable horizontal flow velocity and weir loading rate to minimize turbulence in order to avoid resuspension of settled particles. Typical design values are shown in Table 5.1.

Table 5.1   Basic Design Criteria for
Horizontal Flow Sedimentation Tanks

Parameter	Design value
Surface loading rate $(m^3/m^2 \cdot d)$	20–60
Mean horizontal velocity (m/min)	0.15–0.90
Water depth (m)	3–5
Detention time (min)	120–240
Weir loading rate $(m^3/m \cdot d)$	100–200

Various features must be incorporated into the design to obtain an efficient sedimentation process. The inlet to the tank must provide uniform distribution of flow across the tank. If more than one tank exists, the inlet must provide equal flow to each tank. Baffle walls are often placed at the inlet to distribute even flow, by use of 100–200 mm diameter holes evenly spaced across the width of the wall, as demonstrated in a scale model study reported by Kawamura (1976).

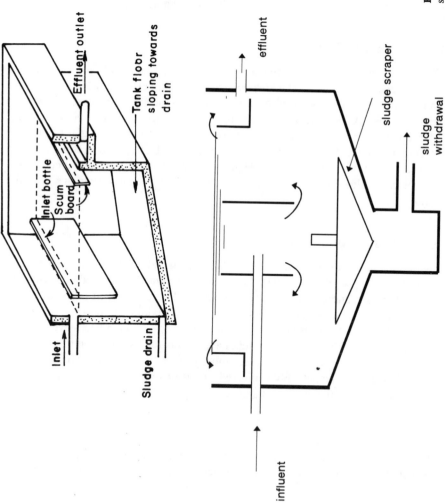

FIGURE 5.1a Schematic diagram of a rectangular sedimentation tank.

FIGURE 5.1b Schematic diagram of a circular sedimentation tank.

The most important factor to be incorporated in the design of the outlet structure is that the particles settled should not be resuspended. The weir length should be sufficient such that the approach velocity is not high enough to create turbulence and resuspend the settled particles and break the floc. Commonly, outlets run the full width of the tank as well as part of its length for rectangular tanks and the full circumference for circular tanks.

Sludge scrapers are employed to remove sludge from the floor of the basin. Mechanical scrapers are usually used with the velocity kept below 0.3 m/min to avoid resuspension, but suction scrapers are also used that can have a velocity of 1.0 m/min. The sludge is then disposed, usually by piping to sludge lagoons for drying. It can later be used as a landfill material.

The design and application status of the conventional sedimentation tanks are discussed in detail in the literature. This chapter is confined to the discussions on tube settlers and the sludge blanket clarifier (clariflocculator).

5.3 TUBE SETTLERS

Due to the long settling time needed, rectangular or circular tanks are large in size and thus occupy extensive land areas. Since the settling efficiency largely depends on the surface area of the clarifier, it is always of concern to increase the settling area available. A successful method to increase the surface area without requiring a large land area is to use tube or plate settlers. This technique has proved quite successful in developed countries and is most cost effective in upgrading existing plants. Since tube settlers do not require any mechanical devices, except for a mechanical sludge scraper in a large installation, they are appropriate as a low-cost option.

5.3.1 Basic Concepts

The basic concept of the tube settler is the same as that of a conventional settling basin. According to Hazen (1904), the amount of sediments removed in a sedimentation basin is proportional to the surface area and is independent of the detention time. In tube settlers, the surface area is increased by dividing the basin by horizontal plates into several compartments, one on top of the other. Some later developments (Culp and Culp, 1974) indicate that tubes and commercially produced tube modules clarify with greater efficiency. Inclined tubes or modules facilitate efficient sludge removal. When mathematically expressed, surface overflow rate V_s is given as the ratio of flow rate (Q) and surface area (A_s):

$$V_s = Q/A_s$$

Therefore, as A_s is increased due to the larger surface area of tubes, or plates, the flow rate can be increased proportionally, keeping the same surface overflow rate.

5.3.2 Different Types

Horizontal Tube Settlers

With horizontal tube settlers, the settling occurs as the water to be clarified flows through the horizontal tube. The tubes are slightly inclined (5°) in the direction of flow, to facilitate easy drainage of sludge during cleaning. The horizontal tube settlers require frequent cleaning to wash down the sludge accumulated at the bottom of the tubes into the sludge storage basin. The backwashing (BW) of the filter is coupled with cleaning and draining of the tubes. When water is drained from the tubes, the falling water scours the sludge down to the sludge storage tank. The last portion of the filter backwash water is used to clean the tubes after the draining is over. Due to the discontinuity in operation during desludging, the horizontal tube settlers are not suitable for plants with a capacity of more than 7 MLD, and for plants requiring minimal or no maintenance. The process is similar to that of inclined tube settlers (see below) in all other aspects. A typical horizontal tube settler is shown in Figure 5.2a.

Inclined Tube Settlers

The most common tube settling devices used are inclined tubes, circular or rectangular in section, packed and fabricated to form a module. A typical inclined tube settler is shown in Figure 5.2b. The module has tubes of small diameter.

The water to be clarified is allowed to flow up through the inclined tube, during which settling occurs. The tubes are inclined at an angle of 45–60° to horizontal. The higher slope facilitates the gravity drainage of the sludge deposited at the bottom of each tube into the sludge compartment for periodic removal. Typical tube settler installations in rectangular and circular basins are shown in Figures 5.3a and b, respectively.

The need for backwashing is eliminated in inclined tube settlers as compared to the horizontal version (Figures 5.2a, b, and 5.4b). But periodic cleaning is generally needed to avoid any build-up of sludge blanket onto the tube bottom, which may cause algae and odor problems. Because of the advantages of shallow settling depth and continuous sludge removal, inclined tube settlers are suited to high capacity installations and installations requiring little maintenance.

Lamella Separator

Most upflow clarifiers incorporate lamellar flocculation elements, one of which is shown in Figure 5.4a. It operates by cocurrent downward flow of both sludge and flocculated water. The classical tilted tubes or lamella are equipped with sludge movement deflecting blades (Figure 5.4b). Even though these blades may cause some turbulence (favorable for flocculation but critical for clarification), they have a general improving effect when placed in the sludge zone (Masschelein, 1992).

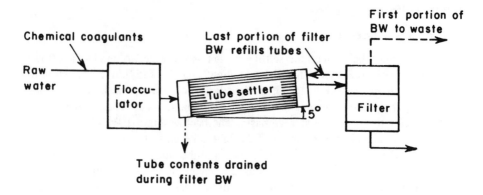

FIGURE 5.2a   Nearly horizontal tube settler. (Weber, 1972. Reprinted by permission of John Wiley & Sons)

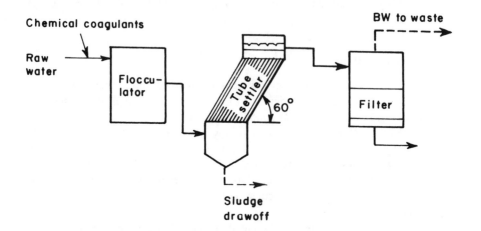

FIGURE 5.2b   Steeply inclined tube settlers. (Weber, 1972. Reprinted by permission of John Wiley & Sons)

5.3.3 Design Criteria

The design of tube settlers basically depends on the following criteria:

1.   Overflow rate, V_s
2.   Shape and size of tube
3.   Length of tube, ℓ
4.   Inclination of tube, Θ

Overflow Rate

Overflow rate is an externally controlled parameter. Yao (1970) has developed the following expression for surface overflow rate:

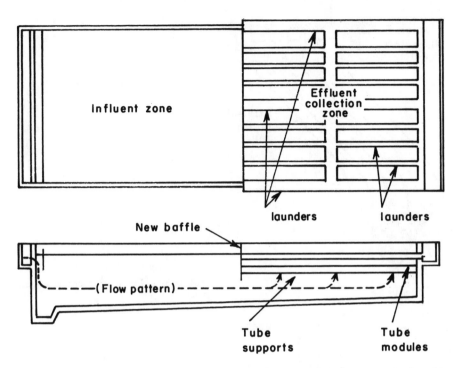

FIGURE 5.3a Typical tube settler installation in a rectangular basin. (Modified from Culp and Culp, 1974)

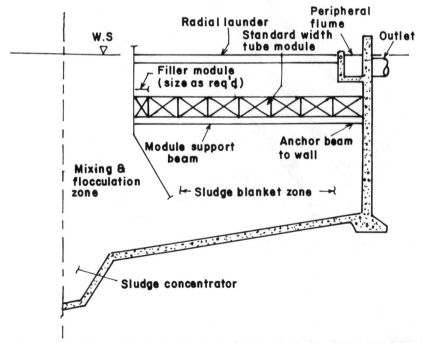

FIGURE 5.3b Typical tube settler installation in a circular basin. (Modified from Culp and Culp, 1974)

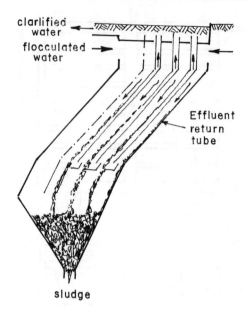

FIGURE 5.4a Lamella separator. (Masschelein, 1992)

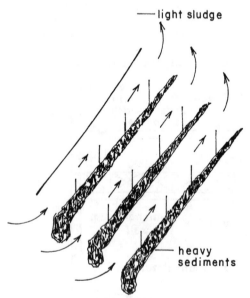

FIGURE 5.4b Countercurrent flow of sludge. (Masschelein, 1992)

$$V_s = C\, S_c V_o\, /\, (\sin \Theta + L \cos \Theta) \qquad (5.1)$$

and

$$V_o = Q/A \qquad (5.2)$$

where V_s = surface overflow rate, $m^3/m^2 \cdot d$; A = area perpendicular to the flow direction m^2; Θ = inclination of tubes to horizontal; L = relative length of tube =

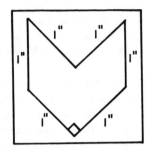

FIGURE 5.5   Cross-section of a Chevron-shaped tube. (Modified from Culp and Culp, 1974)

actual length/diameter of the tube; S_c = critical value of tube settler parameter; V_o = average flowthrough velocity, m/d; Q = flow rate, m^3/d; C = constant, = 6.54×10^5 when V_o is given in fps units and V_s in gpd/ft², = 8.64×10^2 when V_o is given in cm/s and V_s in $m^3/m^2 \cdot d$, = 1 when V_o is given in m/d and V_s in $m^3/m^2 \cdot d$.

Shape and Size of Tube

Tubes are of different shapes such as parallel plates, circular tubes, square tubes, etc., and are available in different sizes. Sizes of circular tubes are expressed in diameter (d), whereas the sizes of plates and square tubes are expressed in depth (d). Specific shapes are available to improve performance. Chevron-shaped tubes, as shown in Figure 5.5, are widely used. The V-groove in the chevron shape enhances the counterflow characteristics of the sludge.

The tube settler parameter S characterizes the performance of the high rate settling system (Yao, 1970). The critical value of S, denoted by S_c is as follows:

S_c = 1 for parallel plates
   = 4/3 for circular tubes
   = 11/8 for square tubes

From equation (5.1), it is clear that surface overflow rate is directly proportional to S_c. Therefore, for a particular V_s, Θ and V_o, the relative length increases with S_c. Hence the specific length will be highest for square tubes and lowest for parallel plates. But parallel plates require larger area. Hence as a compromise, a width to depth ratio of 2:1 to 5:1 is generally used (Yao, 1970). Circular tubes require less L than square tubes, but square tubes require smaller space due to the dead spaces left in between circular tubes.

Length of the Tube

Length of the tube ℓ is the actual length provided by each tube or plate along the flow. But in the design, the relative length L is always used, and it is given by:

$$L = \ell/d \tag{5.3}$$

where d is the diameter of the tube or the perpendicular distance between the parallel plates. Yao (1970) suggests a value of 20 for L to give best performance (from the pilot-scale experimental results). L is a dimensionless parameter.

Rearranging equation (5.1),

$$L = (C V_o S_c - V_s \sin \Theta) / V_s \cos \Theta \qquad (5.4)$$

The above equation is derived based on laminar flow conditions throughout the tube. But in practice a transition zone is developed at the entrance of the tube due to the turbulent flow at this point. Therefore, an allowance for this transition zone must be provided in the tube length by providing an additional length, L'. The following empirical relationship has been proposed by Yao (1970),

$$L' = 0.058 V_o d/v \qquad (5.5)$$

where v is the kinematic viscosity of the fluid.

Therefore, the actual length required = $(L + L')$ d when $L' < L$
$$= 2 Ld \text{ when } L' > L$$

Tikhe (1970) has given the following expression for the minimum relative length. This is based on equation (5.1). The optimum angle of inclination is substituted, which gives a low detention time.

$$L_{min} = [(V_o S_c/V_s)^2 - 1]^{1/2} \qquad (5.6)$$

Willis (1978) put forward an empirical equation for relative length as follows:

$$L \geq 0.787 \times m \times (Q/A) \qquad (5.7)$$

where m = maximum distance in cm from topmost surface to the bottommost surface of the tube, measured perpendicular to the axis.

Inclination of the Tube

The efficiency of a tube settler in removing particles/flocs improves with the increase in tube inclination to the horizontal, Θ, but beyond 35 to 40° the efficiency decreases (Culp et al., 1968). Tikhe (1970) proposed a theoretical equation for minimum angle of inclination that gives minimum detention time, which is obtained from equation (5.1).

$$Q_{min} = \sin^{-1} (V_s/V_o S_c) \qquad (5.8)$$

Detention Time

The lower the detention time, the greater the economy. The detention time (t) can be evaluated from the following equation:

$$t = [(L + L') d/V_o] \qquad (5.9)$$

In fact, the minimum angle of inclination corresponds to the minimum detention time. On the other hand, as indicated earlier, sludge removal will be facilitated by an increase in angle of inclination.

Design of Plenum

Ample space in the plenum (sludge collection chamber) is essential for efficient sludge collection without sludge scouring.

Materials of the Tubes

It is best to use locally available materials. Tubes and plates can be made of plastic, wood, galvanized steel, and asbestos-cement. Each of them has its advantages and limitations. Plastic is suitable in places where a plastic industry is well developed so that the plastic tube modules can be easily fabricated. Asbestos-cement plates must be coated with plastic or a similar type of coating material because of their susceptibility to corrosion from alum treated water. Galvanized steel needs anti-corrosive coatings, and it is not economical to fabricate into tube modules. It is, however, suitable as plates. Wood also needs preservatives and may have to be cleaned often as sludge does not slide easily on wood; it is cumbersome and uneconomical to make tube modules out of wood. Hence, it is better to use plastic in tube modules (of any shape).

5.3.4 Typical Design Values

The loading rates are available through experience. Table 5.2(a) gives some typical loading rates for two ranges of initial turbidities. Table 5.2(b) presents typical values of some design components of tube settlers. Figure 5.6 gives the typical overflow velocities for water of different turbidities. This design criteria diagram could be used to evaluate tube dimensions for particular flow rates and initial turbidities. Final turbidity is considered to be less than 5 NTU.

5.3.5 Application Status

Tube settlers are being used extensively in developed countries and have found important applications in developing countries as well. Their performance is generally found to be superior to those of conventional settling basins. A number of treatment plants using horizontal tubes are in operation with detention times of less than 10 minutes and capacities of 90 to 9000 L/min (Panneerselvam, 1982). A tube settler and mixed media filter in combination with a total detention time of less than 30 minutes provided efficient clarification of highly colored waters, containing filter-clogging algae, iron, and manganese, and raw water with taste and odor, as reported by Culp et al. (1968). Current application includes removal of turbidity (10 to 1000 JTU), color (20 to 200 units), iron (1 to 5 mg/L), manganese, oil, and suspended solids (oil field residue) (Panneerselvam, 1982). Bin (1975) reports that effective removal of all aluminum residuals from the use of alum in the coagulation process was obtained by using anthracite and sand filter with tube settlers.

Table 5.2(a)   Some Typical Loading Rates for Horizontal-Flow Settling Basins Equipped with Inclined Tube or Plate Settlers (Culp and Culp, 1974)

Settling velocity based on total clarifier area	Settling velocity based on portion covered by plates or tubes (m/d)	Probable effluent turbidity (NTU)
Raw water turbidity from 0 to 100 NTU		
120	140	1 to 3
120	170	1 to 5
120	230	3 to 7
170	200	1 to 5
170	230	3 to 7
Raw water turbidity from 100 to 1000 NTU		
120	140	1 to 5
120	170	3 to 7

There are a few limitations in the use of tube settlers, as in any process. The significant limitations are as listed below.

1.   The tube settlers are advantageous when land is expensive. In such cases, conventional settling tanks are advantageous at the start. As the area gets developed, tube settlers could be easily incorporated into the existing conventional clarifiers to cope with the increased loadings. Designing a tube settler at the start itself would require additional new clarifiers during extension to cope with higher loadings later on. This would cost more than the previous option and, hence, is not desirable.

2.   Hot and sunny climates, as found in most countries, cause algal growth on tubes and plates causing maintenance problems.

3.   Careful attention should be paid to the design of inlet and outlet structures. Otherwise, the tube performance will be affected by turbulence and uneven flow through the tubes.

5.4 SLUDGE BLANKET CLARIFIERS

The sludge blanket clarifier incorporates both flocculation and sedimentation in one unit, thereby reducing the plant size. It is also known as a solid contact clarifier or clariflocculator, depending on the design. In a clariflocculator, the flocculation is achieved by mixing the flocculent with turbid water at the central zone of the sludge blanket clarifier and the settling at the outer zone. Various shapes and different designs of sludge blanket clarifiers are available. Sludge blanket clarifiers are also used in

Table 5.2(b)   Typical Values of Design Components of Tube Settlers

Parameter	Value	Reference
Diameter of tube, mm	25–50	Culp and Culp (1974)
Length, cm	60–120	Panneerselvam (1982)
Angle of inclination, degrees	50–60	Yao (1970)
Surface overflow rate, $m^3/m^2 \cdot d$	2.5–9.2	Chen (1979); Willis (1978)

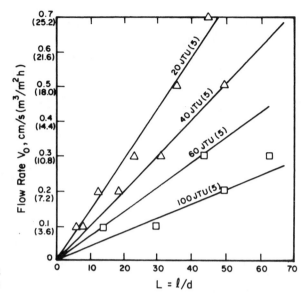

FIGURE 5.6 Design criteria diagram. (Thanh and Chen, 1981)

developing countries with simple modifications. An example is the upflow clari-floc-filter used in India (Dhabadgaonkar and Bhole, 1975). In this section, different types of sludge blanket clarifiers, design concepts, and their operations are reviewed.

5.4.1 Process Description

The water treatment system incorporating a sludge blanket clarifier (SBC) is shown in Figure 5.7. It is seen from Figure 5.7 that the sludge blanket clarifier is immediately followed by filtration. There are several types of SBCs available. Some typical ones are shown in Figure 5.8.

The mixing of flocculent is carried out either hydraulically or mechanically. Sludge is periodically removed from the sludge storage provided at the bottom. When the reactor is started up after some stoppage, it takes time for the blanket to form. To speed up the formation of the sludge blanket, artificially prepared high turbid raw water (e.g., with clay) is sent in (Amirtharajah, 1978). As the hydraulic type does not have any moving parts, it is cheaper to install and maintain. Therefore, it is preferred over mechanical type. Though more expensive, the operation of the mechanical type offers higher flexibility.

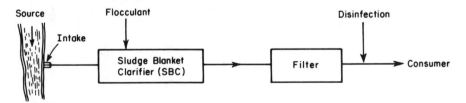

FIGURE 5.7 Water treatment flow diagram with SBC.

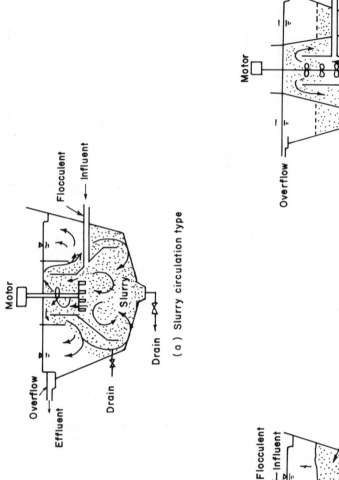

(a) Slurry circulation type

(b) Floc blanket type with paddle flocculation

(c) Combined type

FIGURE 5.8 Clarifiers with sludge blanket flocculation.

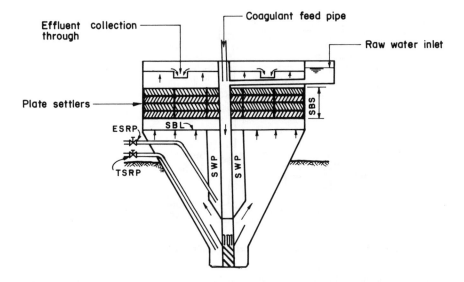

FIGURE 5.9 Modified sludge blanket clarifier. SWP = sludge withdrawal pocket, SBS = sludge blanket stabilizer, SBL = sludge blanket level, ESRP = excess sludge removal pipe, TSRP = thickened sludge removal pipe.

5.4.2 Design Concept

There are theoretical formulations available for the design of SBCs. Miller and West (1966a–c) gives the following basic considerations for the design:

1. Stable blankets are maintained in the column at upflow velocities of up to 5.0 m/h.
2. Increase in blanket depth significantly decreases the floc carry-over. Blanket depth is generally kept between 0.9–2.7 m.
3. Entry velocity of the order of 1.5–1.6 m/h was found to improve the performance of the clarifier.
4. Supernatant depth has no effect on floc entrainment when flow conditions above the blanket are uniform.

It is possible to design SBCs with the above concepts alone. However, pilot-scale studies together with jar tests are recommended. A detailed theoretical formulation with examples is discussed elsewhere (Amirtharajah, 1978).

5.4.3 Application Status

The SBC is quite flexible and can be adopted for use in almost all site conditions. Figure 5.9 shows an interesting SBC system with plate settler arrangement. Here the usual clariflocculator is also equipped with plate settler, thereby facilitating the solid removal. The plate settler sits on top of the clariflocculator. Therefore, in most situations, this unit does not need a subsequent filter unit.

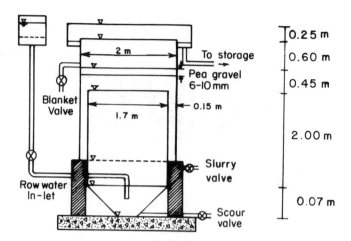

FIGURE 5.10 Sequential solid contact clarifier. (Dhabadgaonkar, 1977)

The operation of this upflow clari-floc-filter can be briefly stated as follows: the raw water enters through a nozzle type inlet into the bottom of the reactor, which is generally in the shape of a cone. The flocculent is dosed at this point and thus gets mixed well due to the turbulence created. The perforated grid arrangement placed at the bottom end of the inlet pipe facilitates even distribution of the water and the flocculent. The water-flocculent mixture gets flocculated while rising toward the top where the sludge blanket is highly concentrated. The flocs are kept in suspension all the time due to the upflow velocity. The incoming particles combine with flocculent to form microflocs, which constantly come into contact with the already formed flocs in suspension. Little by little, the flocs accumulate at the top of the blanket and the surplus falls down over the central sludge withdrawal pockets. The water coming up from the sludge blanket is almost clean. Whatever particles that are left in it also gradually settle down, and the remaining particles are then removed in the plate settler unit.

Another application, again in India, is the sequential solid contact clarifier with hopper bottom that is cylindrical and conical in shape. A pea gravel filter is placed on top of the clariflocculator. It incorporates three separate units (flocculation, sedimentation, and filtration) into one. Figure 5.10 is a schematic diagram of this reactor. This design provides a 2-m-deep sludge blanket and a 15-cm sludge withdrawal zone. A pea gravel (6–10 mm) bed of 0.15-m thickness, suspended 0.3-m above the slurry weir level accelerates the formation of flocs. At the top of this gravel bed, a floc blanket forms due to the slowdown of flow, further helping in the entrapment of particles and flocs. It is called a "sequential solids contact clarifier" because it provides solid contact opportunities in series. The gravel bed filter provides about 30–40% additional turbidity removal (compared to the normal clariflocculators) as observed at Kannan Water Works in India (Dhabadgaonkar, 1977). Some information pertaining to this type of clarifier is given in Table 5.3.

Table 5.3   Parameters of the Sequential Solid Contact
Clarifier (Dhabadgaonkar, 1977)

Parameter	Value
Size of unit	2×2 m $\times$ m
Upflow velocity	1.5–6 m/h
Population served	102–1900 capita
Operation time	18 h/d
Blanket depth	2.0 m
Supernatant depth	50–75 cm
Detention time	45–75 min
Effluent turbidity	less than 5 NTU
Gravel bed cleaning frequency	1–2 months

5.4.4 Relative Merits of the Sludge Blanket Clarifier

Some advantages and disadvantages of SBCs are given in Table 5.4.

Table 5.4   Advantages and Disadvantages of the Sludge Blanket Clarifier

Advantages	Disadvantages
Working of the unit is very simple.	Floc blanket formation after starting up takes time.
Better economy because of the combination of clarification and flocculation and at times filtration as well.	Proper upflow velocity must be provided to prevent sludge settling. Possible clogging of inlet.
Suitable for small community water supplies due to ease of maintenance and operation.	Low turbidity waters require addition of artificial turbidity at start-up for rapid blanket formation.
Can be applied to varying initial turbidities, particularly for high turbidities.	Care must be taken to prevent algal growth.
Head loss is low compared to the other conventional techniques.	Hydraulic overloading or turbidity spikes may lead to floc carryover and plugging of the filters.
	SBC may require a skilled operator, particularly when the raw water quality and the flow rates are highly variable.

5.4.5 Conclusion

The clariflocculator seems to have some clear advantages, even though it looks slightly sophisticated. It is the complicated theory that is sophisticated, but not the reactor itself. Some established detail designs of the SBC are available and could easily be incorporated into any new designs. There are modified versions incorporating a plate settler and filter to achieve entire solid-liquid separation in the same unit itself. Most of these modifications are made to suit the need of developing nations. Its use is highly recommended in developing countries — especially in small community water supply schemes — because of its flexibility in capacity and its ability to take up widely varying turbidity loads. It is highly suitable for package treatment plants, which are useful in remote areas as well as in congested urban areas.

EXAMPLE 5.1

Design a sedimentation tank for a flow (Q) of 1000 m³/d (see Figure 5.11). Determine the dimensions of the tank and the outflow weir length. Assume suitable design criteria.

Solution

Assume an overflow rate (OFR) of 20 m³/m²·d.

$$\text{Area} = \frac{Q}{\text{OFR}} = \frac{1000}{20} = 50 \text{ m}^2$$

Assuming a detention time (DT) of 2 h,

$$\text{Volume} = Q \times DT = 1000 \times \frac{2}{24} = 83.3 \text{ m}^3$$

$$\text{Depth} = \frac{V}{A} = \frac{83.3}{50} = 1.7 \text{ m}$$

If width (W) to length (L) ratio is 1:3, then

A = 3W² = 50
W = 4.1 m
L = 3W = 12.3 m

Assuming a weir loading rate (WLR) of 160 m³/m·d,

$$\text{Weir length} = \frac{Q}{\text{WLR}} = \frac{1000}{160} = 6.3 \text{ m}$$

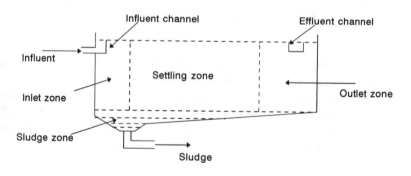

FIGURE 5.11

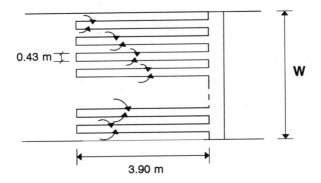

FIGURE 5.12   Overflow weir arrangement in the sedimentation tank.

Example 5.2

Design a rectangular sedimentation tank for a flow rate (Q) of 25,000 m³/d with the following data:

- Detention time (DT) = 3 h
- Length (L) to width (W) ratio = 4:1
- Surface loading rate = 20 m³/m²·d
- Bottom slope = 1:100
- Weir loading rate = 0.1 m³/m/min
- The overflow weir to be designed will have the geometry shown in Figure 5.12.

Solution

Required volume of the tank:

$$V = DT \times Q = 3 \times \frac{25,000 \text{ m}^3/d}{24} = 3125 \text{ m}^3$$

$$\text{surface loading rate} = 20 \text{ m}^3/\text{m}^2 \cdot d = \frac{Q}{W \times L} = \frac{25,000}{4W^2} \quad (\because L = 4W)$$

$$W = \sqrt{\frac{25,000 \text{ m}^3/d}{4 \times 20 \text{ m}^3/\text{m}^2 d}} = 17.70 \text{ m}$$

$$L = 4 \times W = 4 \times 17.7 = 70.8 \text{ m} \approx 71 \text{ m}$$

$$V = L \times W \times H = 3125 \text{ m}^3$$

$$\therefore H = \frac{3125}{17.7 \times 71} = 2.50 \text{ m}$$

Adopted dimensions of sedimentation tank are

> L = 71 m
> W = 17.70 m
> H = 2.50 m + 0.5 m freeboard

Weir Design

$$\text{Total weir length needed} \;=\; \frac{\dfrac{25,000\ \text{m}^3/\text{d}}{24 \times 60\ \text{min}}}{0.1\ \text{m}^3/\text{m} \cdot \text{min}} = 173.6\ \text{m}$$

Available width = 17.70 m
Further weir length needed = 173.6 – 17.7 = 155.9 m
Required number of finger weirs of length 3.9 m each = 155.9/(3.9 × 2)
$$= 19.98$$
$$= 20$$

EXAMPLE 5.3

Design a (coagulation-sedimentation) settling tank with a continuous flow for treating water for a population of 45,000 persons with an average daily consumption of 135 L/person. Assume a detention period of 4 hours. (See Figure 5.13.)

Solution

Average consumption = 135 × 45,000 = 6,075,000 L/d.

Allow 1.8 times for maximum daily consumption: maximum daily consumption = 1.8 × 6,075,000 = 10,935,000 L/d. Therefore, capacity of the tank = (10,935,000/1000) × (4/24) = 1822.5 m³.

Assume depth of tank = 3.5 m. Therefore, surface area of the tank = 1822.5/3.5 = 520.7 m².

Assume width of tank = 12 m. Therefore, length of the tank = 520.7/12 = 43.4 m. Therefore, adopt a tank of size = 43.5 m × 12 m × 3.5 m.

Assuming a depth of 0.3 m for sludge. Depth of shallow end (at the effluent end) = 3.5 + 0.3 = 3.8 m.

Assuming a bottom slope of 1 in 60. Depth of the deep end (at the influent end) = 3.8 + (1/60) × 43.5 = 4.525 m. At the entry to the tank, a floc chamber should be provided.

Assume the capacity of the floc chamber is 1/16 of the settling chamber. Capacity of floc chamber = 1822.5/16 = 113.9 m³. If the depth of floc chamber is 2.5 m, then the area of the floc chamber = 113.9/2.5 = 45.56 m². The flocculation chamber also has a width equal to the sedimentation chamber, i.e., 12 m. Therefore, length of floc chamber = 45.56/12 = 3.796 ≈ 3.8 m.

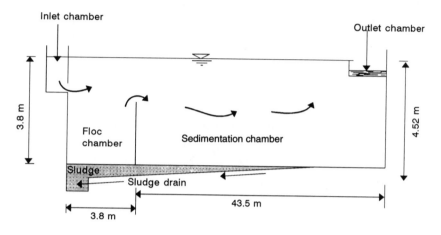

FIGURE 5.13

EXAMPLE 5.4

Design a horizontal tube settler with square tube arrangements for a flow (Q) of 100 m³/d, given that

- Surface overflow rate (V_s) = 36 m³/m²·d = 36 m/d
- Cross-sectional velocity (V_0) = 0.25 cm/s
- Thickness of tube settler = 5 cm

Solution

The relative length of the tube settler:

$$L = \frac{CV_0 S_c - V_s \sin\theta}{V_s \cos\theta} = \frac{8.64 \times 10^2 \times 0.25 \times \frac{11}{8}}{36} = 8.25$$

Correction factor for inlet turbulence:

$$L' = \frac{0.058\, V_0 d}{v} = \frac{0.058 \times 0.25 \times 5}{10^{-2}} = 7.25$$

Since $L > L'$, relative length $= L + L' = 15.5$. Thus, the length of tubes $= 15.5 \times 5 = 77.5$ cm.

Cross-sectional area of the tubes:

$$A = \frac{Q}{V_0} = \frac{100 \text{ m}^3/\text{d}}{0.25 \frac{\text{cm}}{\text{s}} \times \frac{1 \text{ m}}{10^2 \text{ cm}} \times \frac{24 \times 60 \times 60 \text{ s}}{1 \text{d}}} = 0.463 \text{ m}^2$$

Adopt 0.6 m width and 0.8 m depth. Total number of tubes required $= 0.6 \times 0.8/0.05^2 = 192$.

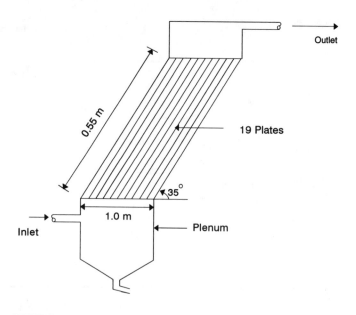

FIGURE 5.14

EXAMPLE 5.5

Design an inclined plate settler for a flow of 1 MLD using the following design values. (See Figure 5.14.)

- Surface overflow rate (V_s) = 48 m³/m²·d.
- Cross-sectional velocity (V_0) = 240 m³/m²·d = (240 × 100) / (1440 × 60) cm/s = 0.28 cm/s.
- Shape factor constant for parallel plates (S_c) = 1.
- Spacing between plates = 5 cm.
- Kinematic viscosity (v) = 1.002 × 10⁻² m²/s.

Solution

Plate Inclination

Minimum plate inclination is given by:

$$\theta_{min} = \sin^{-1}\left(\frac{V_s}{V_0 S_c}\right) = \sin^{-1}\left(\frac{48}{240 \times 1}\right) = 11.5°$$

Usually, θ is 30–50°. Select a plate inclination (θ) of 35°.

Length of the Plate Settler

The relative length of the plate settler:

$$L = \frac{CV_0S_c - V_s \sin\theta}{V_s \cos\theta} = \frac{8.64 \times 10^2 \times 0.28 - 48\sin 35}{48\cos 35} = 5.45$$

Check for L_{min}

$$L_{min} = \left[\left(\frac{V_0S_c}{V_s}\right)^2 - 1\right]^{\frac{1}{2}} = 4.9 < 5.45$$

Correction factor for inlet turbulence,

$$L' = \frac{0.058\, V_0 d}{v} = \frac{0.058 \times 0.28 \times 5}{1.002 \times 10^{-2}} = 8.1$$

Since $L' > L$, total relative length $= 2L$. Actual length of the plate settler: $L_a = (2Ld)$ $= (10.9 \times 5) = 54.5$ cm. (Note: Generally, tube length ranges from 60–120 cm.)

Cross-Sectional Area

$$A = \frac{Q}{V_0} = \frac{1000}{10 \times 24} = 4.17\ \text{m}^2$$

Adopt 1 m width and 4.17 m depth. Number of plates required $= (100/5) - 1 = 19$.

EXAMPLE 5.6

Design a solid-contact clarifier for a design flow rate of 0.166 m³/s with the following design data. (See Figure 5.15.)

Flocculation time = 0.5 h
Flocculator depth = 3.0 m
Distance of paddles from the shaft = 65–75% of the radius of the flocculator
Paddle velocity = 2 RPM
Effective area of paddles = 10% of sweep area of the flocculator
Relative velocity of paddles with respect to water = 0.25 × paddle velocity
Detention time of the sedimentation tank = 3 h

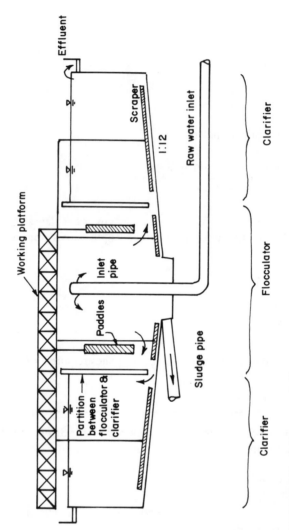

FIGURE 5.15   Solid contact clarifier.

Solution

Flocculator Tank Dimensions

Flocculator volume = 0.166 m³/s × 30 × 60 s
    = 298.8 m³
Surface area of flocculator = (298.8/3)
    = 99.6 m²
Flocculator chamber diameter = $[(4 \times 99.6)/\pi]^{1/2}$
    = 11.26 m
Sweep area = 11.26 m × 3 m
    = 33.78 m²
Paddle area = 10% of sweep area (Bhole, A. G., personal communication)
    = (10/100) × 33.78
    = 3.378 m²

Assume there are four paddles and length of each paddle is 2.5 m,

Therefore, width of paddles = [3.378/(4 × 2.5)]
    = 0.34 m

Assume paddles are 3.5 m away from the shaft (about 65–75% of radius of flocculator)

Then rotation speed of paddle = [(2π × 3.5 × 2)/60]
    = 0.73 m/s

Power Requirements of Flocculator

Power input and velocity gradient can be calculated using the following equations:

$$P = \left[1/2\,C_D\,A_p\,\rho\,v_1^3\right]$$

$$G = \left[(P/\mu V)^{1/2}\right]$$

where A_p = area of paddles = 3.38 m², C_D = Newton's drag coefficient (1.8), G = Velocity gradient, P = Power input, V = Volume of flocculator = 298.8 m³, v_1 = Relative velocity of paddles with respect to water, = 0.75 × 0.73 = 0.547 m/s, μ = Water viscosity = 10^{-3} kg·s/m², ρ = Water density = 1000 kg/m³.

Therefore, power input (P) = [1/2 C_D A_p ρ v_1^3]
    = 1/2 × 1.8 × 3.38 × 1000 × (0.547)³
    = 497.88 kg m/s

Velocity gradient (G) = $[(P/\mu V)^{1/2}]$
    = [497.88/(10^{-3} × 298.8)]
    = 40.82 s⁻¹

Sedimentation Tank Dimensions

Area of sedimentation tank = 0.166 m³/s × 3 × 60 × 60 s = 1792.8 m³. Assume effective depth of sedimentation tank (excluding sludge zone) is 3 m. Therefore, sedimentation tank area = 1792.8/3 = 597.6 m². If diameter of sedimentation tank is D, then $[(\pi D^2/4) - \pi (11.26)^2/4] = 597.6$ m². Therefore, diameter (D) = 29.79 m. Assuming 20% depth for sludge accumulation and a 10% slope, then the depth of sedimentation tank at the center point

$$= 3 + 3 \times (20/100) + (29.79/2) \times (10/100)$$
$$= 3 + 0.6 + 1.49$$
$$= 5.09 \text{ m}$$

EXAMPLE 5.7

Design a sludge blanket clarifier of a type given in Figure 5.16 for a flow of 1 MLD given an influent concentration of 100 mg/L and the following design criteria.

1. Solids retention time = t_s = 100 h.
2. The product $G\phi t$ = 4600, ($G\phi t$ = contact opportunity).
3. Solids concentration in sludge blanket (SB) = 20,000 mg/L.
4. The ratio of total depth (L) to sludge blanket depth (ℓ) = 2.

G, ϕ, and t are velocity gradient, floc volume fraction, and retention time, respectively. Here t refers to the sludge retention time (Vigneswaran and Dharmappa, 1992).

Solution

Determine the Surface Area (A)

Floc volume fraction = ϕ = 20,000 × 10⁻⁶ = 0.02
Solids retention time = t_s = 100 h = 3.6 × 10⁵ s
The contact opportunity of particles ($G\phi t$) = 4600
G = 4600/0.02 × 3.6 × 10⁵ = 0.6389 s⁻¹

The velocity gradient (G) is related to power requirement:

$$G = (P/\mu V)^{1/2}$$
$$P = \rho (\Delta H) Q g \text{ watts}$$

where Q = flow rate, ΔH = headloss through the sludge blanket, g = acceleration due to gravity, ρ = density of water, V = volume of the solid contact clarifier = Qt_w, t_w = hydraulic detention time.

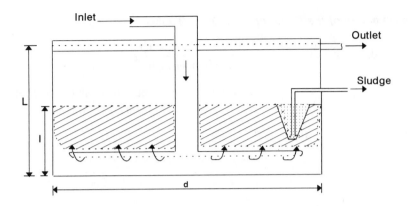

FIGURE 5.16 Schematic of sludge blanket clarifier.

Substituting for P and V in the equation for G:

$$G = [\rho(\Delta H)g/\mu t_w]^{1/2} = [1(\Delta H)\ 981/(1.002 \times 10^{-2}\ t_w)]^{1/2}$$

$$\Delta H = 4.169 \times 10^{-6}\ t_w$$

if one equates the headloss to the buoyant weight of sludge blanket, the headloss across the sludge blanket = (effective density) × (floc fraction) × (height of SB)

$$\Delta H = (1.005 - 1) \times (20,000 \times 10^{-6})\ \ell = 1 \times 10^{-4}\ \ell$$

Equating the above two equations,

$$\ell = 4.169 \times 10^{-2}\ t_w$$

From the definition of hydraulic retention time (t_w),

$$t_w = V/Q = AL/Q = 2Al/Q \quad (\text{since } L = 21)$$

$$A = t_w\ Q/21$$

Substituting for t_w and l,

$$A = \left\{\Delta H\ Q/\left(4.169 \times 10^{-6}\right)\right\}/\left\{\left(\Delta H/2 \times 10^{-4}\right)\right\}$$

$$= (Q/0.08338) = \left(1.157 \times 10^{4}\right)/0.08338 = 13.88 \times 10^{4}\ cm^{2} = 13.88\ m^{2}$$

Provide a circular clarifier of diameter (d) = 4.2 m.

Determination of the Depth of the Sludge Blanket

$$\text{Solids retention time} = \frac{\text{weight of solids in tank}}{\text{rate of solids entering into the tank}}$$

$$= \frac{\text{solids volume} \times \text{concentration of sludge blanket}}{\text{flow rate} \times \text{influent concentration}}$$

$$100 = A\ell(20,000)/100 \, Q$$

$$\ell = \frac{Q \times 100 \times 100}{A \times 20,000} = \frac{41.67 \times 100 \times 100}{13.88 \times 20,000} = 1.5 \text{ m}$$

Determination of the Depth of the Clarifier

$$L = 2 \, \ell = 2 \times 1.5 = 3 \text{ m}$$

REFERENCES

Amirtharajah, A., Design of rapid mix units, *Water Treatment Plant Design for Practicing Engineers*, R. L. Sanks (Ed.), Ann Arbor Science, Michigan, 1978.

Bhole, A. G., *Modified Sludge Blanket Clarifiers for Better Water Quality in Rural Areas*, Indian Water Works Association, Jan–March, 83, 1993.

Bin, N. C., *Optimization of the Inclined Tube-Settler and Anthracite Sand Filter*, AIT Master's Thesis No. 834, Asian Institute of Technology, Bangkok, 1975.

Chen, Y. R., *Design Criteria for Inclined Tube Settlers*, AIT Master's Thesis No. EV-79-29, Asian Institute of Technology, Bangkok, 1979.

Culp, G. L., Hansen, S. and Richardson, G., High rate sedimentation in water works, *J. AWWA*, 60, 681, 1968.

Culp, G. L. and Culp, R. L., *New Concepts in Water Purification*, Van Nostrand Reinhold, New York, 1974.

Dhabadgaonkar, S. M., *Package Sequential Solids Contact Clarifiers for Low Cost Drinking Water Production in Rural Areas from Surface Sources*, Vysvesvaraya Regional College of Engineering, Nagpur, India, 1977.

Hazen, A., On sedimentation, *Transactions of the American Society of Agricultural Engineers*, 53, 45, 1904.

Kawamura, S., Considerations on improving flocculation, *J. AWWA*, 68, 328, 1976.

Masschelein, W. J., Lamella and tubular assisted settling processes, *Unit Processes in Drinking Water Treatment*, Marcel Dekker, New York, 1992.

Miller, D. G. and West, J. T., Floc blanket clarification, I, *Water and Water Engineering*, 70, 240, 1966a.

Miller, D. G. and West, J. T., Floc blanket clarification, II, *Water and Water Engineering*, 70, 291, 1966b.

Miller, D. G. and West, J. T., Floc blanket clarification, III, *Water and Water Engineering*, 70, 341, 1966c.

Panneerselvam, K. N., *Use of Tube Settlers for Tropical Surface Water Treatment*, AIT Master's Thesis, Asian Institute of Technology, Bangkok, 1982.

Thanh, N. C. and Chen, Y. R., Importance of relative length in tube settlers design, *Water Resources Bulletin — J. AWRA*, 1981.

Tikhe, M. L., Some theoretical aspects of tube settlers, *Indian J. Environ. Health*, 16(1), 26, 1970.

Vigneswaran S. and Dharmappa H. B., *Unpublished Notes on Wastes and Wastewater Treatment*, University of Technology, Sydney, 1992.

Vigneswaran, S., Shanmuganantha, S. and Mamoon, A. A., Trends in water treatment technologies, *Environmental Sanitation Reviews,* Environmental Sanitation Information Center, Asian Institute of Technology, Bangkok, No 23/24, 1987.

Weber, W. J., *Physicochemical Processes for Water Quality Control*, Wiley Interscience, 1972.

Willis, R. M., Tubular settlers — A technical review, *J. AWWA*, 70, 331, 1978.

Yao, K. M., Design of high rate settlers, *Proc. ASCE, Env. Eng. Div.*, 99(EE5), 621, 1970.

Yao, K. M., Theoretical study of high rate sedimentation, *Journal of Water Pollution Control Federation*, 42, 218, 1970.

6: Filtration

CONTENTS

6.1 INTRODUCTION

Filtration technologies are classified under two major categories, namely slow sand filtration and rapid sand filtration, depending mainly on the mode of filtration. Slow sand filters, which include biological activity in addition to physical and chemical mechanisms for removing impurities from the raw water, are especially suitable for small community water supplies, because of their large areal requirement. Numerous documents are available on this technology. A brief review of the slow sand filter is presented in Section 6.2 of this chapter.

Slow sand filtration is suitable only for low turbid waters. Most of the surface waters are of high turbidity. Appropriate filtration and prefiltration technologies adopted in small community water supplies for turbid raw waters are discussed in Section 6.3.

The rapid sand filter, on the other hand, due to its lower areal requirement (25 to 150 times less than the slow sand filter), is used widely as a final clarification unit in municipal water treatment plants. There have been several modifications made on the conventional rapid sand filters to improve the filter design. These are discussed in the Section 6.4 of this chapter.

Section 6.5 of this chapter discusses the direct filtration process, which is becoming important in treating the low turbid surface waters.

6.2 SLOW SAND FILTER

In rural areas, especially in developing countries where land is plentiful, the slow sand filter can be used with success if the raw water is not highly turbid. It is well suited for turbidities less than 50 NTU. In case of highly turbid water, one has to have

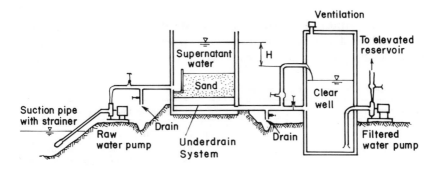

FIGURE 6.1 Simplified operation of slow sand filter. (Huisman, 1978)

pretreatment prior to slow sand filtration. The schematic diagram of a simplified slow sand filtration plant, serving 1,500–20,000 people, is presented in Figure 6.1.

6.2.1 Principle and Operation

Water is purified by passing it through a bed of fine sand at low velocities (0.1–0.3 $m^3/m^2 \cdot h$), which causes the retention of suspended matter in the upper 0.5–2 cm of the filter bed. By scraping out this top layer, the filter is cleaned and restored to its original capacity. The interval between two successive cleanings varies from a few weeks to a few months, depending on the raw water characteristics.

The removal mechanisms of particles in a slow sand filter include mechanical straining, sedimentation, diffusion, and chemical and biological oxidation. Coarse and fine particles of suspended matter are deposited at the surface of the filter bed by the action of mechanical straining and sedimentation, respectively, while colloidal and dissolved impurities are removed by the action of diffusion. By chemical and biological oxidation, the deposited organic matter is converted into inorganic solids and discharged with filter effluent. Microbial and biochemical processes, and hence the removal of impurities, take place mainly in the top zoological layer of the filter bed (known as the "Schmutzdecke").

The important design parameters of a slow sand filter are the depth of the filter bed, filter media size, the filtration rate, and the depth of the supernatant water level. As far as possible, these design parameters should be based on experience obtained from existing treatment plants that use a raw water source of similar quality. In the absence of reliable data from such existing treatment plants, pilot plants can be used to determine suitable design criteria. The values presented in Table 6.1 can serve as helpful guidelines.

Since the purification mechanism in a slow sand filter is essentially a biological process, its efficiency depends upon a balanced biological community in the "Schmutzdecke." Therefore, it is desirable to design the filters to operate as far as possible at a constant rate. However, the operation of slow sand filters in most developing countries is intermittent due to the financial difficulties in employing operators to run the plants around the clock. But the intermittent operation causes deterioration in effluent quality because during stoppages the microorganisms causing bacteriological degradation of trapped impurities lose their effectiveness. Intermittent operation disturbs the continuity needed for efficient biological activity.

Table 6.1 Design Summary of a Slow Sand Filter

Design parameters	Recommended range of values
Filtration rate	0.15 m³/m²·h (0.1–0.2 m³/m²·h)
Area per filter bed	Less than 200 m² (in small community water supplies to facilitate manual filter cleaning)
Number of filter beds	Minimum of two beds
Depth of filter bed	1 m (minimum of 0.7 m of sand depth)
Filter media	Effective size (ES) = 0.15–0.35 mm
	Uniformity coefficient (UC) = 2–3
Height of supernatant water	0.7–1 m (maximum 1.5 m)
Underdrain system	
Standard bricks	
Precast concrete slabs	Generally no need for further hydraulic calculations
Precast concrete blocks with holes on the top	
Porous concrete	
Perforated pipes (laterals and manifold type)	Maximum velocity in the manifolds and in laterals = 0.3 m/s
	Spacing between laterals = 1.5 m
	Spacing of holes in laterals = 0.15 m
	Size of holes in laterals = 3 mm

One way of overcoming this problem is by allowing the filter to operate at a declining rate after a cycle of constant rate filtration. Declining rate filtration gives an additional water production of 0.5 and 0.7 m³/m² (of filter area), with a declining rate operation for 8 or 16 hours after 16 or 8 hours of constant rate, operation respectively. The effluent achieved during this operation is generally satisfactory. Moreover, the declining-rate mode may be applied during night-time, resulting in significant savings of labor.

6.2.2 Cleaning

Before cleaning a slow sand filter, floating matter such as leaves and algae that may cause nuisance should first be removed by raising the water level in the unit to flush the floating matter over the weir. Then the water level in the bed is lowered to about 0.1–0.2 m below the sand surface by closing the inlet valve and opening both the supernatant water drain valve and the valve on the underdrains. The filter bed is then cleaned by rapidly scraping off the top 1–2 cm of the bed, in view of minimizing the interference with the filter microbial activity. When one unit is shut down for cleaning, the others are run at a slightly higher rate to maintain the output of the plant.

After cleaning, the unit is refilled with water through the underdrains. This water can be obtained from an overhead storage tank or by using filtered water from an adjacent filter. The temporary reduction of plant output due to this method should be taken into account when the clear-water storage tank is designed, assuring that sufficient water is available for the users.

Before starting operation, the filter needs a period of at least 24 hours to allow for reripening of the bed. After this period, the microorganisms usually reestablish to be able to produce an acceptable effluent. In cooler areas, the ripening may take a few days. Even then, if the turbidity of the effluent is sufficiently low, the water supply can be resumed after a period of one day, after adequate chlorination (Visscher, 1990).

6.2.3 Costs

The construction cost of an open slow sand filter excluding pipes and valves has two components, namely, (1) the floor, underdrains, sand, and gravel; and (2) the walls of the filter box. The cost of labor and land is usually less important. In rural areas, the cost of land rarely exceeds 1% of the total construction cost. However, in densely populated areas, the somewhat larger areas required for slow sand filter plants can be a problem (Visscher, 1990).

The construction cost of small- and medium-sized slow sand filters are often less than that of other types of treatment. Construction of a slow sand filter in India is less expensive than a rapid sand filter up to a capacity of 3000 m³/d. However, if recurrent costs for operation and maintenance are taken into account, the balance shifts to 8000 m³/d. Using cheaper materials like ferrocement and less expensive drainage systems will reduce the cost of slow sand filtration (Visscher, 1990).

In Colombia, the new slow sand filtration plants are competitive with conventional treatment plants. Even conversion of conventional plants into slow sand filtration plants has been cost-effective. For example, in a suburb near Cali, operation and maintenance of a conventional treatment plant cost U.S. $1240/month. Because of the increasing cost of chemicals, it was decided to reconstruct the plant and adopt slow sand filtration as the main treatment. Most of the existing structure could be used, holding the cost of reconstruction to as low as U.S. $7000. The oversized water storage tanks were changed into slow sand filters. Operation and maintenance costs are now U.S. $240/month (Visscher, 1990).

A cost estimate (Paramasivam et al., 1981) based on 1979 prices (in Nagpur, India, excluding the contractor's profit) has shown that the filter bed cost per m² is Rs 350 (U.S. $43.75), and the cost per meter of wall length is Rs 570 (US $71.25). They reported that there is no cost penalty for building more than two filters for gaining reliability and flexibility in operation. This study revealed that the number of filters can be raised from two to five by spending roughly 6–22% more money.

6.2.4 Advantages and Disadvantages

Simplicity of design and operation and minimal requirements of power and expensive chemicals make the slow sand filter an appropriate technique for the removal of organic and inorganic suspended matter as well as pathogenic organisms present in surface waters of rural areas in developing countries. Sludge handling problems are also minimal. Close control by an operator is not necessary, which is important to small communities because an operator may have several responsibilities. Design experience in America shows that the slow sand filter is more than 99.9% efficient in removing *Giardia* cysts and coliform bacteria, and provides a stable effluent quality with a low operating budget (Seelaus et al., 1986). It has the added advantage of being able to make use of locally available materials and labor.

However, slow sand filtration has certain limitations. The requirements of a large area, large quantities of filter medium, and labor for the manual cleaning are the major disadvantages. These limitations do not apply to the rural areas in the developing world where land and unskilled labor are readily available.

6.3   SIMPLE FILTRATION AND PRETREATMENT FACILITIES

In rural areas in developing countries, there is no problem of land availability. Therefore, one could use the slow sand filters with success if the raw water is not highly turbid (<50 NTU). If the turbidity is high, the raw water has to be pretreated prior to slow sand filtration.

In place of slow sand filters, one could also use different types of filtration systems depending on the size of the community, water source, soil condition below the water source, etc. This section discusses the different types of filtration technologies that could be used in the rural areas in developing countries.

6.3.1   Horizontal Flow Coarse-Media Filtration

Horizontal flow coarse-media filtration using coarse gravel or crushed stones as filter media is a sound prefiltration system in handling turbid water (>50 NTU). The main advantage of horizontal flow filtration is that when raw water flows through it, a combination of filtration and gravity settling takes place, which invariably reduces the concentration of suspended solids.

The horizontal prefilter design follows the rectangular sedimentation tank with inlet, outlet, and filtration/sedimentation zones (Figure 6.2). In the direction of flow, water passes through various layers of graded coarse material in the coarse-fine-coarse sequence. Each layer of gravel is separated by a strong wire mesh.

Design Criteria

The range of values of important design parameters of horizontal flow prefilters are summarized in Table 6.2.

Applications

This type of filter has been successfully used in a few small community water supply schemes in Thailand as prefilters prior to slow sand filters. One such installation in Jedee-Thong village in Thailand indicates that this prefilter is capable of removing

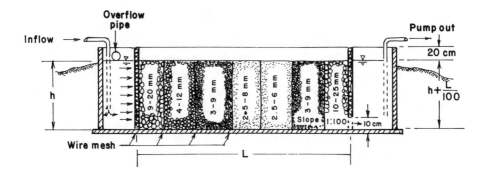

FIGURE 6.2   Horizontal flow coarse-media filter. (Thanh and Hettiaratchi, 1982)

60–70% of the suspended solids and 80% of the coliform organisms in raw water
(Thanh and Hettiaratchi, 1982)

Table 6.2   Design Parameters of Horizontal Flow Pre-Filter

Parameter	Range of values
Filtration rate	0.3–1.0 m³/m²h
Optimum filtration rate	0.5 m³/m²h for low turbid waters (15–50 NTU)
	0.3 m³/m²h for high turbid waters (up to 150 NTU)
Depth of filter bed	1 m (0.8–1.5 m)
Water level	0.8 m
Free board	0.2 m
Length of the filter	5 m (4–10 m)
Length to width ratio	1:1–6:1
Area occupied by the filter	10–100 m²
Specification of filter bed	9–20 mm gravel
(in the direction of flow)	4–12 mm gravel
	3–9 mm gravel
	2.5–8 mm gravel
	2.5–6 mm gravel
	3–9 mm gravel
	10–25 mm gravel
Slope at the bottom	A slope of 1/100 is provided toward the effluent end to facilitate the flow of pretreated water
Covering of inlet and outlet compartments	Required if the filter is exposed to sunlight

Advantages and Disadvantages

This type of horizontal prefilter can successfully be used as a pretreatment facility for
turbid waters.

- Capital and installment cost of this system is low.
- Operation and maintenance can be done using unskilled labor.
- Cleaning of this prefilter is not frequent.
- It can withstand the seasonal variations of raw water quality.
- Land requirement for filter construction is high.

Cost

The installation in Jedee-Thong village indicates that the total capital cost for construction of the prefilter, slow sand filter (3 compartments), weir chambers, and clear well and installation of water distribution system for a design population of 1000 is only U.S. $6947 (i.e., US $7 per capita). The operational and maintenance cost calculations indicate that the cost of a cubic meter of water produced is approximately U.S. $0.05 only, which is reasonable for a rural water supply. (These cost calculations were based on the Thailand price index in 1977 and unit cost of water was based on operation and maintenance cost only.)

6.3.2 Two-Stage Filter (Coconut Fiber/Burnt Rice Husk Filter System)

Frankel (1979) developed an appropriate technology filter that consists of two stages:

- the first-stage filter with shredded coconut husk fiber as filter medium (roughing filter) to filter out the coarse suspended solids from the water
- the second-stage filter with burnt rice husk as filter medium (polishing filter) to remove the residual turbidity and other contaminants

A typical design of a two-stage filter plant is shown in Figure 6.3. The detailed study on two-stage filtration has shown that the combination of shredded coconut fiber and burnt rice husk (in series) will effectively remove particulate matter from water and also to some extent remove certain dissolved materials.

Design Criteria

The media characteristics and operating parameters of both coconut fiber and burnt rice husk filter are given in Table 6.3.

Applications

This two-stage filtration can be used for turbid waters (up to 150 NTU) in an economical way.

Advantages

- This two-stage process operates at 10–15 times higher filtration rates than the slow sand filter, which reduces the filter construction cost considerably.
- Removes color and taste due to the adsorption capability of filter media.
- It is cost-effective (several installations in Southeast Asia indicate that the total construction cost, including pump, filters, storage jars, and public taps, amounted to less than US $2 per capita).

- Simple in construction, operation and maintenance.
- It can handle high turbidities.

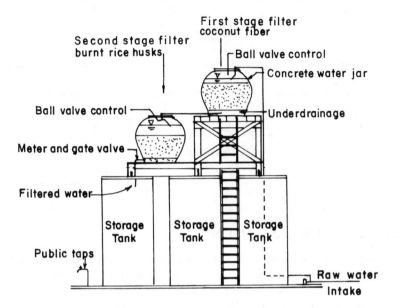

FIGURE 6.3 Two-stage filtration unit constructed at Ban Sam, Changwat Korat, Thailand. (Frankel, 1979)

Table 6.3 Design Parameters of Two-Stage Filtration Unit

Parameter	Coconut fiber filter (first-stage filter)	Burnt rice husk filter (second-stage filter)
Filter depth	60–80 cm	60–80 cm
Free board	1.0 m	1.0 m
Filtration rate	1.2–1.5 m³/m²h	1.2–1.5 m³/m²h
Underdrain system		
Gravel layer (supporting media)	Pea gravel of 3–6 mm (5–10 cm depth)	
Lateral, manifold underdrain system	Main drain and lateral pipe material: GI or PVC	
	Spacing between orifices: 0.3 m	
	Spacing between laterals: 0.3 m	
	Diameter of an orifice: 0.6 cm	
	Ratio of area of orifice to lateral: 1:2	
	Ratio of area of lateral to main drain: 1:1.5	

Disadvantages

- In most cases bacteriological removal is insufficient, which requires some simple postdisinfection.
- After a few months of operation, due to the biodegradability and odor problem, the coconut fiber had to be replaced with fresh material.

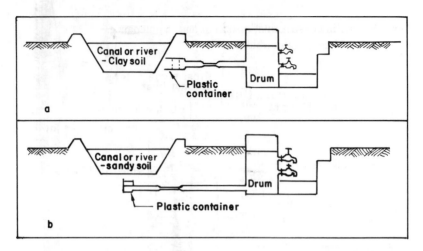

FIGURE 6.4 Modified shore filtration. (Jahn, 1981)

6.3.3 Modified Shore Filtration

The principle of the modified shore filtration treatment method is the same as that of an infiltration gallery. Here, cylindrical plastic pipes with small holes are located under the canal or spring bed. The location of pipe depends on the canal's or spring's soil condition and the flow rate. If the soil is clay-type, then the pipe should be placed horizontally. If it is sandy, then the pipe should be placed vertically (Figure 6.4). An inverted filter screens the water flow into the pipe. The filter media specifications are given in Table 6.4. The water from the canal or spring bed is allowed to trickle through the holes in the top surface of the pipe and flow through it. This water is later collected in a drum placed near the canal or spring bed.

Design Criteria

The values of important design parameters are listed in Table 6.4.

Table 6.4 Design Parameters of Shore Filtration

Parameter	Values
Pipe	PVC, diameter 40 cm
Filter media	Sand, 0.3–0.5 mm
	Gravel, 0.7–0.9 mm
	(Gravel of larger sizes also can be used)

Limitations

This type of modified shore filtration unit functions only for part of the year (during the dry season or flood season, this unit cannot be operated). Thus, the consumers cannot depend on this unit throughout the year for their water supply.

6.3.4 Seawater Supplies (SWS) Filter System

An SWS unit developed in the United Kingdom applies the new concept of filtration of raw water right at its source, as in the case of modified shore filtration. The SWS unit is not strictly considered as a filter, but it is a device to use the sea bed or river bed itself as an efficient natural filter. This type of unit was initially used for marine use, but it can also be used for potable water supply.

Unit Description

The SWS unit is a rectangular box with a false ceiling consisting of a compression-molded slotted plate. The unit is built of corrosion-free fiberglass. The small community-scale SWS unit (cross-section 60 × 30 cm) weighs about 8 kg. This unit is buried, open end down, in the sea or river bed as shown in Figure 6.5. The top of the unit should be at least 15 cm below the river bed. From the bottom of the unit, 3/4 of the height is packed with coarse sand and gravel media. The remaining 1/4 of the height of the unit is empty, and a suction pipe is connected to this part of the unit. Sand is then filled around the unit in such a way that there will not be any empty space around the unit. The suction pipe of the unit is connected to the intake pump (Cansdale, 1979a,b). When this pump is switched on, water from the bottom of the water bed trickles through the river bed and reaches the filter media of the unit. From here, the water is filtered by the filter media as the water flows up to the top of the unit. The clean water is then taken out through the suction pipe.

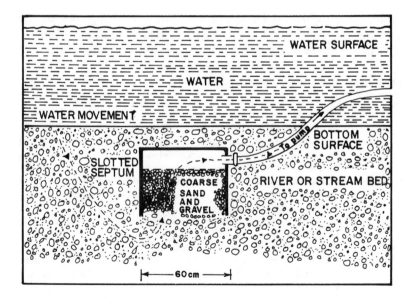

FIGURE 6.5 The SSW filter system.

Design Criteria

The design criteria of the SWS filter are presented in Table 6.5.

Advantages

This filter system has many advantages as listed below:

- No moving parts and no chemicals are needed, so maintenance is minimal.
- Simple process, no need for skilled operators.
- Can be operated continuously or intermittently.
- For industries, this unit can be used as a pretreatment unit; therefore, one can avoid other pretreatment operations such as sedimentation.
- Space requirement is minimal.
- Remarkably low cost system (for a large unit with an output of 40,000 L/h, the total cost over a period of 5 years was £360 (in 1979) in the United Kingdom).

Disadvantages

- This unit can work only where there is permanent surface water.
- In deep muddy beds, steep rocky shores, and gorges, this unit cannot be used.

Applications

This filter system has been used in Malaysia, the Philippines, Singapore, and Thailand to supply water for fish farming, swimming pools, industrial use, agricultural use, etc. This system can also be used as a prefiltering unit, which will reduce significantly the

Table 6.5   Design Parameters of SWS Filter System

Parameter	Range of values
Unit	1. Square-shaped box of 60 × 60 cm (maximum) with a height of 40 cm.
	2. A hollow section of 10 cm between the top portion of the media to roof of the box (this hollow section has a series of distance pieces to give stability and stop the collapse of the unit under vacuum).
Filter bed media	Coarse sand and gravel of size between 0.5 and 5.0 mm can be used (both very fine sand, especially of windblown origins (0.2 mm), and very large stones (50 mm) are not suitable as filter media).
Pipe size	Resistance (friction head) in the pipe increases rapidly with both rate of flow and reduction in pipe size.
Suction head	The total suction head should be kept below 7 m.
Delivery line	Flexible armored hose is needed from the unit to (at least) highest water level. Semirigid PVC pipes can be used from the river edge and also for the rest of the delivery line.
Pumps	Hand pump with 4 m³/h capacity (only operational cost of a hand pump is replacement of new diaphragm for every 400 hours running).

subsequent treatment. A filter system of this type has been installed in India to treat the highly turbid river Ganges water for potable water supply (Nigam, 1981).

6.3.5 Conclusion

As summarized in Table 6.6, the selection of suitable filtration system or prefiltration system depends on the raw water characteristics, pretreatment, post-treatment, land availability, availability of materials, skilled labor, size of community, soil conditions, etc. When more than one of the above-mentioned filtration systems are suitable technically, the cost and the degree of operation and maintenance become the selection criteria.

6.4 RAPID FILTERS

6.4.1 Design Criteria of Rapid Sand Filters

Rapid filtration is used as the final clarifying step in municipal water treatment plants. If the raw water has turbidity in excess of 10–20 NTU, the rapid filters must be provided with efficient flocculation and sedimentation as pretreatment units. In rapid filters there is practically no biological action; at the most there is some nitrification in certain cases when the speed is limited, when the oxygen content is adequate, and when the nitrification bacteria find favorable nutritive conditions in the water.

There are two types of rapid sand filters, namely, the gravity and pressure types. Table 6.7 gives a comparison of the characteristics of the rapid filters (both pressure and gravity types) with slow sand filters. A diagrammatic section of a rapid gravity filter is given in Figure 6.6.

The working of a rapid gravity filter is explained below with reference to Figure 6.7. When the filter is in a working condition, only values 1 and 4 are kept open and all others are kept closed (see Figure 6.6).

6.4.2 Backwashing

When the water passes through the filter medium, supporting layer, and underdrain, it experiences frictional loss of resistance known as headloss. When the headloss exceeds 1.5–2.5 m, the filter needs cleaning. The first operation is to close valves 1 and 4 (Figure 6.6) and allow the filter to drain until the water lies a few centimeters above the top of the bed. Then valve 5 (Figures 6.6 and 6.7) is opened, and air is blown back through a compressed air unit at a rate of about 1 to 1.5 m^3 free air/min-m^2 of bed area for about 2–3 minutes, at a pressure of 20–35 kN/m^2. The water over the bed quickly becomes very dirty as the air-agitated sand breaks up surface scum and dirt. Following this, valves 2 and 6 are opened, and an upward flow of water is sent through the bed at a carefully designed high velocity sufficient to expand the bed (20–50% expansion) and cause the sand grain to be agitated so that deposits are washed off them, but not so high that the sand grains are carried away in the rising upflush of water.

Table 6.6 Summary of Alternative Filtration Methods for Small Community Water Supplies

Parameter	Modified shore filtration	Slow sand filtration	Two-stage filtration	SWS filtration	Horizontal flow prefilter
Raw water requirement	Not specified	<50 NTU	<150 JTU	Can be used even for high turbid waters	<200 NTU
Extent of treatment	Used as pretreatment unit	Water of low turbidity and good bacteriological quality is produced	Water of low turbidity and good bacteriological quality is produced	Used as a pretreatment unit	Used as a pretreatment unit
Pretreatment	—	If the raw water turbidity is more than 50 NTU, then a pretreatment unit such as horizontal flow filter is necessary	If the raw water turbidity is greater than 150 NTU, then a multi-stage unit or a coagulation step is necessary	—	—
Post-treatment	Depends on water quality, slow sand filter followed by disinfection	Preferably disinfection	Preferably disinfection	Depends on raw water quality (generally slow sand filter followed by (disinfection)	Depends on raw water quality (generally slow sand filter followed by disinfection)
Filtration rate	—	$0.1–0.2\ \mathrm{m^3/m^2h}$	$1.2–1.5\ \mathrm{m^3/m^2h}$	—	$0.5–2\ \mathrm{m^3/m^2h}$
Filter media size	Sand (covering material is optional)	Sand ES = 0.45–0.55 mm, UC = 2–3	Coconut fiber shredded and washed. Burnt rice husks of ES = 0.3–0.5 m, UC = 2.3–2.6	Coarse sand and gravel of size 0.5–5 mm can be used	Gravel compartments of 9–20 mm 4–12 mm 3–9 mm 2.5–8 mm 2.5–6 mm 3–9 mm 10–25 mm

Table 6.6 Continued

Parameter	Modified shore filtration	Slow sand filtration	Two-stage filtration	SWS filtration	Horizontal flow prefilter
Filter media depth	—	1–1.4 m	0.6–0.8 m	0.3 m	0.8 m
Underdrain system	—	Lateral manifold system (for small filter units) or standard bricks or precast concrete blocks with holes in the top or porous concrete (this underdrain system is under the supporting gravel layer)	Lateral manifold system under the supporting gravel layer	—	—
Supernatant level	—	1–1.5 m	1.0 m	0.1 m	—
Cleaning procedure	—	Scrape out the top few centimeters and replace with new sand (or wash scraped sand and reuse)	Used coconut fiber is replaced with new (or washed) coconut fiber material; in the case of burnt rice husk, top 10 cm is scraped off and replaced with new media	—	Cleaning is done periodically, compartment by compartment, gravel in the particular compartment is taken out and cleaned and and placed again
Cleaning frequency	—	Once every 2 months (frequency depends on raw water quality)	Once in every 3 to 4 months (depending on raw water quality)	—	Once in 6 months to 1 year (depends on raw water quality)

Table 6.6   Continued

Parameter	Modified shore filtration	Slow sand filtration	Two-stage filtration	SWS filtration	Horizontal flow prefilter
Construction					
Cost of construction	Average	Average (if the appropriate sand size is available)	Low	Low	Average
Land requirement	Large	Large	Average	Small	Average
Materials of construction	Wall supports of brick, concrete block in weak soil	Concrete, ferrocement, or reinforced concrete	Wooden support structure, concrete or GI jars for filter tank	Rectangular fiberglass box of $0.6 \times 0.6 \times 0.4$ m	Concrete structure
Ease of construction	Difficult	Less difficult	Simple	Simple	Less difficult
Skilled operator requirement	No	No	No	No	No
Operation					
Ease of operation	Simple	Simple	Simple	Simple	Simple
Maintenance cost	Low	Average	Low	Low	Low
Special requirement	Near the river or water source	None	None	Near the river or water source	None
Size of community per unit	Unlimited	Medium population	Less than 1000 people	Large unit with an output up to 40,000 L/h, about 6000 people served	Medium population

Table 6.7 Comparison of Slow Sand Filters and Rapid Sand Filters

Characteristic	Slow sand filter	Rapid filter	
		Gravity	Pressure
Filtration rate	2–5 m³/m²·d	120–360 m³/m²·d	
Size of bed	Large (2000 m²)	Small (100 m²)	
Depth of bed	300 mm gravel, 1 m sand, unstratified	500 mm gravel, 0.7–1.0 m sand, stratified; in some cases sand and anthracite are used as dual media	
Effective size of sand	0.35 mm	0.6–1.2 mm	
Uniformity coefficient	2–2.5	1.5–1.7	
Head loss	Up to 1 m	Up to 3 m	
Length of run	20–90 days	1–2 days	
Method of cleaning	Scrape off top layer and wash (or replace with new sand)	Backwash with water and air + water scour and in some cases surface scour	
Washwater consumption	0.2–0.6% of filtered water	3–6% of filtered water	
Penetration of suspended solids through the filter bed	Superficial	Deep	
Pretreatment by coagulation	No	Yes	Yes
Covered construction	No	Optional	Yes
Visible operation	Yes	Yes	No
Capital cost	High	High	Medium
Operating cost	Low	High	High
Skilled supervision	Not required	Required	
Adjusting the quality of filtrate	Difficult	Can be done quickly	
Bacteria removal	99.99%	90–99%	

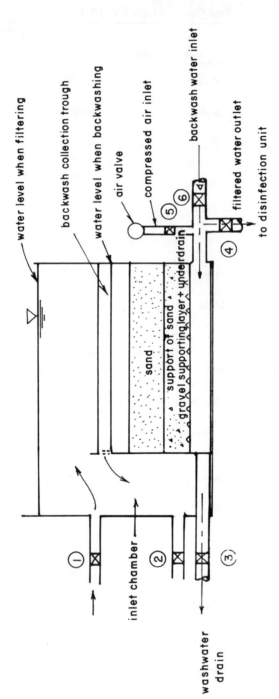

FIGURE 6.6 Diagrammatic section of a rapid sand filter.

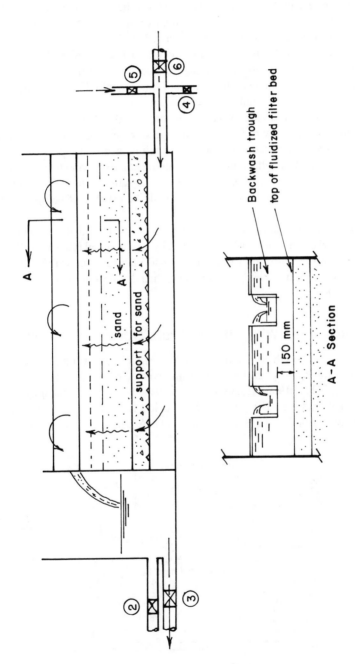

FIGURE 6.7   Diagrammatic section of a backwash system.

After the washing of the filters has been completed, valves 2 and 6 will be closed, and valves 1 and 3 are opened. This restores the inlet supplied through the valve 1. The filtered water is wasted to the gutter for a few minutes after this, until the required quality is achieved. Ultimately, valve 3 is closed and valve 4 is opened to get the filtered water again. The entire process of backwashing the filters and restarting the supplies takes about 15 minutes. The specified minimum backwash time for a rapid filter is 5 minutes. The amount of water required to wash a rapid filter may vary from 3–6% of the total amount of water filtered.

Upward washwater rates are usually of the order of 0.3–1.0 m/min. It is usually more economical to use gravity flow from a large storage tank, since the backwash flow rates are high. Energy requirements can be minimized by keeping such a tank topped up by a relatively small, continuously running pump drawing from the filtered water supply. The washwater tank should have a capacity to provide a minimum of one filter wash of 10 minutes duration. It should be capable of being refilled in 60 minutes.

If there is significant particle retention at the surface of the bed, it is recommended to have a surface wash. Surface wash is achieved by an additional 0.06–1.2 m³/min·m² of water jetted onto the surface of sand at 150–450 kN/m², as shown in Figure 6.8. Different filter backwash methods and the recommended design values are summarized in Table 6.8.

6.4.3 Backwashing With Effluent From Other Filter Units

In this method of backwash, which results in "interfilter backwashing" of filters, the filter units of a treatment plant have to be interconnected as shown in Figure 6.9 (Valencia, 1977). The effluent from the interconnected units is collected by a single drainage channel, and the filter effluent outlet is located at a higher level than that of the washwater troughs of the individual units. A particular unit is cleaned by

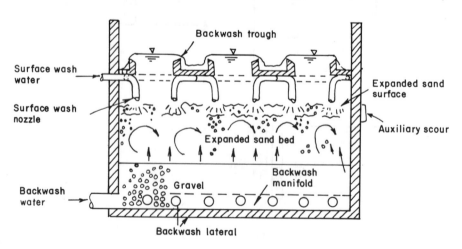

FIGURE 6.8   Diagrammatic presentation of surface wash.

Table 6.8 Recommended Design Values for Various Backwash Methods (Adapted from Vigneswaran et al., 1983)

Parameter	High-rate water backwash	Low-rate water backwash		Water backwash with air auxiliary		Water backwash with surface wash auxiliary
		Air scour followed by low-rate water backwash	Simultaneous air and low-rate water backwash followed by low-rate water backwash	Air scour followed by high-rate water backwash	Simultaneous air and low-rate water backwash followed by high-rate water backwash	
Backwash rate	37.5 m³/m²·h	18 m³/m²·h	15–18 m³/m²·h	>18 m³/m²·h	—	15–18 m³/m²·h
Backwash water pressure	2.5–5 kg/cm²	2.5–5 kg/cm²	2.5–5 kg/cm²	2.5–5 kg/cm²	2.5–5 kg/cm²	0.25–0.5 kg/cm²
Air scour rate	—	27 m³/m²·h	18–27 m³/m²·h	27 m³/m²·h	36–46 m³/m²·h	—
Surface wash rate	—	—	—	—	—	10–12 m³/m²·h
Pressure of surface scour water	—	—	—	—	—	1.5–4 kg/cm²
						This type of filter backwash is used when mud ball formation occurs on the top of filter bed
Porosity range during expansion	0.68–0.70	—	—	—	—	—
Expansion of medium	80–100%	Low	Low	Low	Low	—
Time of washing	3–6 min	3–6 min	2–3 min	3–4 min	2–3 min	—
Time of air scour application	—	3–6 min	2–3 min	3–4 min	2–3 min	—
Amount of wash water needed	High	Low	Low	High	High	High
Efficiency of cleaning action	Poor	Fair	Good	Good	Good	—
Applicability	Single and multimedia filters	Single media only	Single media only	Single and multimedia filters	Single and multimedia filters	Single media filters

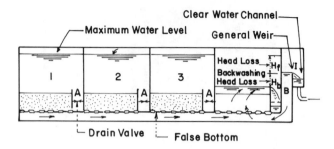

FIGURE 6.9 Backwashing of one filter with the flow of the others. (Valencia, 1977)

closing the inlet and opening the drainage outlet of that unit. The water level in the unit is thus lowered, and a positive head (H_b) is created, which reverses the direction of flow through the filter bed. The backwashing will start with the effluent from the adjoining units providing the upward flow. Once the filter-cleaning operation is finished, the outlet (drainage) is closed, and the inlet of the unit is opened to resume the filtration process.

Schulz and Okun (1983) briefly describe the following design principles. The head available for backwashing (H_b) is the difference in elevation between the effluent weir and the gullet lip of the filter. To obtain sufficient head for backwashing, the depth of the filter box must be substantially greater than for conventional filters. The filter box for the plant in Cali, Colombia, has a depth of 6 m, compared with a depth of about 3 m in conventional filters. The headloss is only 20 to 30 cm compared with about 1 to 1.4 m for conventional filters.

The filter bottom can be produced from concrete beams with plastic tube orifices inserted along the beam. The spacing of orifices dictate the headloss in the system. As long as the underdrain systems are interconnected, the backwash velocities will be low enough in the plenum so that wash water distribution will be fairly uniform. Therefore, by increasing the depth of the water over the filter beds to about 1.5 to 2.5 m, limiting the headloss in the underdrain system to about 20 to 30 cm, interconnecting the underdrain system, and using dual-media filter beds, the backwashing headloss can be made sufficient to produce the desired expansion of the filter media.

Interfilter washing filtration units can be designed either with unrestricted or restricted declining flow rate. The following precautions must be taken into account in designing this mode of backwash.

- For one filter to be washed with the flow of others, the total production of the plant must be at least equal to the washwater flow needed to clean one filter.
- The filter units must supply enough water for the required backwash rates. A minimum of four filter units, capable of working at a rate one-third higher, is necessary to minimize the peak flow requirement when one unit is out of service for washing.

- The filters must be so designed that one may be taken out of service for repairs without interruption of the normal service of the others.
- The underdrain must be specially designed to produce low headloss. This is feasible because the filters are completely open at the bottom, and the washwater flow rate is, therefore, very low.

This system can be used for both single and multimedia filters, but it requires four or more filter units in order to operate efficiently. Filter backwashing with this method has been practiced in Australia for a long time, and has been successfully used in more than 100 installations in the United States. Filters of this kind have also been operating satisfactorily in large plants in Latin America, including those serving Mexico City (24 m³/s); Monterrey, Mexico (24 m³/s); Rio Grande, Brazil (6 m³/s); and Cali, Colombia (4 m³/s); as well as in Peru, Bolivia, and the Dominican Republic.

6.4.4 Filter Underdrains

The underdrain system of a filter is an important component in the design and operation. The selection is based on the filter type and size, media characteristics, and the selected method of backwashing. The underdrain should contain a central manifold, with laterals that are either perforated or have umbrella-type strainers on top. Other types that may be used include clay tile blocks, precast concrete laterals, false bottoms with strainers, and porous plates, etc. The selected underdrain system, while ensuring uniform flow distribution of the backwash, should ensure durability, reliability, and cost-effectiveness.

There are two methods of keeping a uniform flow distribution of the backwash water inside the filter cell:

1. By making the orifice or slits in the filter underdrain system small enough to introduce a controlled headloss
2. By decreasing the flow velocity of the pressurized conduit upstream of the underdrain system so that the hydraulic grade and energy lines of the flow entering the underdrain system are fairly uniform

The false-bottom underdrain systems are preferred due to reliable performance and low maintenance costs (Kawamura, 1991). Table 6.9 gives the typical design parameters of an underdrain system (lateral manifold system) for a small rapid sand filter.

6.4.5 Improvements on Rapid Filter

Improvements on Filter Media

The conventional rapid filter generally uses sand with an effective size of 0.6–1.2 mm and a uniformity coefficient of 1.5–2. After the backwash, the filter media gets

Table 6.9 Design Criteria for Underdrains (NWS & DB, 1988)

Criterion	Value
Minimum diameter of underdrains	20 cm
Diameter of the perforations	6–12 mm (staggered at a slight angle to the vertical axis of the pipe)
Spacing of perforations along laterals	7.5 cm for 6 mm perforations
	20 cm for 12 mm perforations
Ratio of total area of perforation to total cross-sectional area of laterals	0.25 for 6 mm perforations
	0.50 for 12 mm perforations
Ratio of total area of perforation to the entire filter area	0.003
Length to diameter ratio of the lateral	60:1
Maximum spacing of laterals	30 cm
Cross-sectional area of the manifold	1.5–2.0 times the total area of laterals
Velocity of the filtered water outlet	1.0–1.8 m/s

stratified with the finer medium remaining at the top and the coarser medium at the bottom of the filter bed. This makes only the top fine sand portion effective in filtration. To overcome this problem, two alternatives have been proposed: dual or multimedia filtration; or coarse-size, narrowly graded media filtration.

For dual-media filtration, the size and specific gravity are carefully selected to minimize intermixing. The commonly used media are anthracite coal and sand. Various researchers have given optimum size ratios to avoid intermixing (Table 6.10). Extensive research has been carried out to study the advantages and disadvantages of intermixing of filter media in dual media filtration. While some researchers feel that the grain size of coarse anthracite and fine sand should be chosen in such a way that the intermixing at the interface is minimized, others believe that controlled mixing among filter media is beneficial. A detailed design of media is discussed in the literature (Mazumdar, 1984; Vigneswaran et al., 1983).

Table 6.10 Calculated Size Ratios to Avoid Mixing (Vigneswaran and Ben Aim, 1989)

Calculated effective size ratio (anthracite:sand)	Comment	Reference
2.38:1	Neglects settling, based on laminar flow backwash	Fair et al. (1968)
2.63:1	Neglects backwashing, considers hindered settling	Conley and Hsiung (1969)
2.73:1	Neglects backwashing, considers hindered settling	McCabe and Smith (1967)
3.00:1	Based on experience	Camp et al. (1971)
2.07:1	Based on pilot-scale study	Mazumdar (1984)

Another alternative is to replace a graded single medium with a coarse, narrowly graded medium of larger depth. This arrangement results in deeper penetration of the suspended solids and thus higher storage capacities, which would lead to a longer run. The selection of size depends on the raw water and required effluent qualities. The commonly used size range is 0.9–1.1 mm. This arrangement can meet an increase in demand in existing units because it can be operated at a higher filtration rate. However, deeper penetration of the solids would entail a higher backwashing requirement. The rate of air and water used for backwashing depends on the size of the medium. For example, sand of 2 mm effective size requires a washing rate of 90–110 $m^3/m^2 \cdot h$ of air and 19–24 $m^3/m^2 \cdot h$ of water.

As a result of the search for a dual-media filter in which interface intermixing does not occur, a new type of filter called the flotofilter has been introduced. The flotofilter uses synthetic plastic beads as the filter medium. Polypropylene is the coarse medium and polystyrene is the fine medium (Werellagama, 1993). Both the media have a specific gravity less than water, hence the bed floats in water. Polypropylene is much more dense than polystyrene, hence it forms the lower part of the floating filter bed. As the coarse medium is at the bottom, the filtration is upflow. This filter needs an upper retention grid to prevent the loss of media with filtered water. Due to the large density difference, the two media do not intermix even under severe agitation, leaving a clear media interface after backwashing. This eliminates the most common operational problem encountered in conventional multimedia filters. Also, as there is a buffer zone of water between the floating bed and the filter inlet, the bed is not disturbed due to the turbulent conditions in the underdrain system. This permits the use of a very simple underdrain system, resulting in both capital and operational cost savings. Since the two media are floating, they can be fluidized using much lower energy than conventional dual-media filters. This results in considerable energy and water saving during the backwashing of the filter. Experiments carried out in France (Ben Aim et al., 1993) and in Australia (Vigneswaran and Ngo, 1993) have indicated that this can be successfully used as a static flocculator and/or prefiltration unit prior to the direct filtration unit.

Improvements on Flow Rate

Conventional rapid filtration operates at a constant rate of approximately 5 $m^3/m^2 \cdot h$. Research work on the variation of flow rate has indicated that high-rate filtration and declining-rate filtration are advantageous for most cases. If it is possible to achieve the desired filtrate quality with a higher filter rate at an operational and maintenance cost comparable to that of a conventional rapid filter, then one could achieve a significant capital saving by using a high-rate filter. Similarly, research has indicated that declining-rate filtration produces a better effluent quality than the conventional process. Therefore, if one could obtain the same amount of filtered water with equal capital investments, then declining-rate filtration would have a definite advantage over conventional rapid filtration. Another advantage of using declining-rate filtration is that it does not require automatic rate control.

The concept of declining-rate filtration is not new. Basically, no rate controller is used in this system, and instead it is replaced by a fixed orifice. The filtration rate in this system is allowed to decline from a maximum value at the beginning of the run, when the filter is clean, to a minimum value at the end of the filter run, when the filter is in need of backwashing. In practice, several (a minimum of four) filters are used in parallel, and the water level is maintained essentially at the same level in all operating filters at all times. This is achieved by providing a relatively large

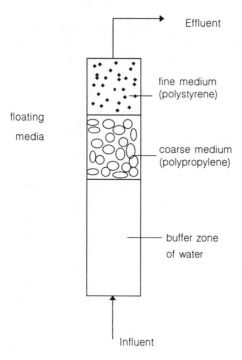

Effluent

fine medium
(polystyrene)

floating

media

coarse medium
(polypropylene)

buffer zone
of water

Influent   **FIGURE 6.10**   The flotofilter.

influent header pipe or channel common to all the filters with a relatively large influent valve or gate to each individual filter. Details on the declining-rate filter operational principles, design criteria, and plant operations can be found elsewhere (Amirtharajah, 1978).

High-rate filtration, in which the filtration rate is about 10–20 $m^3/m^2 \cdot h$, as compared to the rate of conventional rapid sand filtration which is of the order of 5.0 $m^3/m^2 \cdot h$, is useful in upgrading existing plants. Such high filtration rates are possible, thanks to the development of (1) dual-media or multimedia; and (2) control of flocculation by polyelectrolytes, which produce relatively less sludge. High-rate filters with dual or coarse-medium arrangements have been successfully used in developed countries. This could be one of the economic solutions for the expansion of existing water treatment plants and for the construction of new plants. The filters should be designed to operate at the highest practical rate. The design should be such that even with the higher frequency of washing, they are more economical. Here, more attention should be paid to the selection of the filter media and filtration rate so that the filtrate quality meets the required standard.

Modifications of Flow Direction

The flow in conventional rapid filtration is in the downward direction. The major disadvantage in this process is that the flow meets the fine sand before reaching the coarse sand that is found at the bottom layer. If one inverts this flow to an upward direction, then a better use of the filter media can be achieved. But the upflow filter creates a problem of fluidization, especially toward the end of the filter run. This problem can be overcome by using a grid system, as shown in the Figure 6.11, near the top of the filter medium or by operating the filters both in an upward and downward flow direction simultaneously (biflow arrangement).

During filtration the sand arches against the grid, and hence this grid provides sufficient resistance to prevent expansion of the bed and prevents breakthrough or channeling at relatively high rates of upflow filtration. The spacing of the vertical plates should be close enough to provide resistance to sand expansion during upflow filtration, but not so close as to m ake upward flow backwashing difficult. Okun (1967) suggests that the grid spacing should be 100–150 times the size of the fine sand at the top of the bed.

A top layer of anthracite can be used to surround the grid when a high filtration rate is desired. This is because the anthracite seems to provide better arching conditions for preventing bed expansion (Hamann and McKinny, 1968). This design has the advantage of operating at a higher filtration rate of 12.5–25 m^3/m^2h. In this system, backwashing water as a percentage of water produced is also found to be less. This type of upflow grid filers has been extensively used for filtration of water for municipal and industrial supplies in Europe, Africa, and America (Landis, 1966; Okun, 1967).

6.5 DIRECT FILTRATION

Conventional water treatment plants generally use unit operations such as rapid mixing, flocculation, sedimentation, filtration, and disinfection. Depending on the quality of the water, one or more unit operations can be eliminated, thereby achieving a cost-effective water treatment. Direct filtration is one such method. Filters used in direct filtration thus differ little from those for conventional treatment in construction. The primary difference in the operation of the two systems is related to solids storage capacity and backwashing requirements. Different direct filtration flow schemes are presented in Figure 6.12.

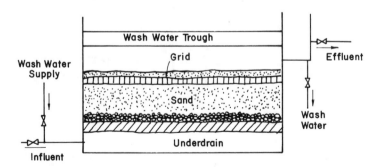

FIGURE 6.11 Upflow filter with a grid system comprising parallel vertical plates.

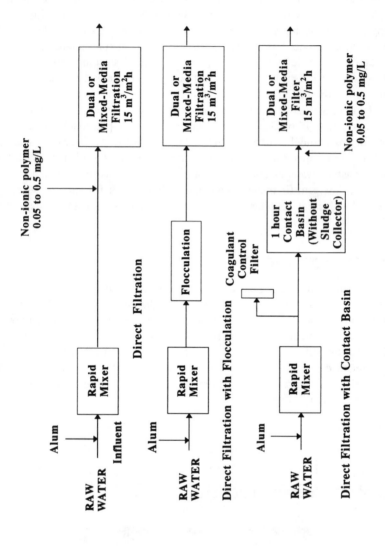

FIGURE 6.12 Direct filtration flow schemes.

Direct filtration was first explored during the early 1900s, but these attempts were not successful, due to the rapid clogging of the sand beds. The development of coarse-sand filters and dual-media filters has made it possible to store greater amounts of floc within the filter bed without excessive headloss, and has thus increased the feasibility of the direct filtration process. Further advances in filter design and the availability of a wider selection of chemical coagulants and poly-electrolytes have resulted in a variety of filtration systems being designed in which coagulating chemicals are employed. The flocculation basin is either eliminated or reduced in size, and the sedimentation basin is not utilized. Such processes thus have only screening, rapid mixing, and a short time of flocculation prior to filtration. All suspended solids and flocs formed are deposited in the filter, which is usually a multi-media, granular bed containing coal, sand, and perhaps other media.

The American Water Works Association (AWWA) Filtration Committee's report (1980) on a worldwide survey of 70 operating and pilot plants has indicated that waters with less than 40 units of color, turbidity consistently below 5 NTU, iron and manganese concentrations less than 0.3 mg/L and 0.05 mg/L, respectively, and algal counts of up to 2000 per mL (measured in absorption units at 1000 nm) appear to be perfect candidates for direct filtration. Turbidity and color removals are consistently attained in this process. By efficient postchlorination, bacteria and virus removal problems can be eliminated. Most of the literature favors the use of dual or mixed media for direct filtration.

Direct filtration can successfully be used for low-turbid waters, because of its lower capital and operational cost. It does not require any sophisticated equipment, although skilled operators are needed in order to monitor the filters. Attention should be paid to the possibility of poor bacteriological quality of the filters in case of highly polluted raw water.

6.5.1 Design Principles

Since optimum design parameters depend greatly on the nature of the water to be treated, pilot studies are required to determine the appropriate type of coagulant and coagulant-aid, and the media composition, size, and depth. Some guidelines are given below.

Filter media: Can be single, dual, or mixed media, but usually dual and mixed media are preferred in direct filtration. Gadkari et al. (1980) recommend dual-media with bituminous coal or anthracite coal. Here deeper filters with greater filter medium depths are preferred (King and Amy, 1979). The finest media possible should be selected to minimize chemical dosages. Within reasonable limits, coarse filter beds can produce the same quality filtrate as finer beds, but more polymer is required. Fine filter media are supported on a gravel bed. This is preferred to direct support on bottoms equipped with mechanical strainers or nozzles, which are not recommended (Culp, 1977).

- Rapid mixing: The rapid mixing process for direct filtration usually does not differ much from that used for conventional plants.

- Filtration rate: 10–15 $m^3/m^2 \cdot h$ (constant-rate operation). A rate of 12 $m^3/m^2 \cdot h$ is usually adopted. Recent studies have shown that higher filtration rates are possible (up to 20 $m^3/m^2 \cdot h$) with low turbid waters (Murray and Roddy, 1993a).

- Flocculents: The type of flocculent is the most important parameter and should be experimentally evaluated initially. Alum has been utilized with success in many installations. Dosages of 5–10 mg/L can be effective for direct filtration of waters of various turbidities (Adin et al., 1979). However, more recently it has become apparent that a carefully selected cationic polymer may have considerable advantages in some situations. Polyelectrolyte doses for uncolored water may frequently vary from 0.05–1.0 mg/L when used as primary flocculents (essentially cationic), and may be less when used in addition to alum for the treatment of organic contaminated water (Adin et al., 1979). The general range of cationic polymer dosage used as the primary coagulant is 0.1–5 mg/L (Culp, 1977). A detailed study conducted in Sydney, Australia (Murray and Roddy, 1993b) indicated that cationic polymer together with alum will not only increase flow rate, but also significantly increases the filter run time. A nonionic polymer filter aid was found to play an important role in maximizing filter performance. Lag time between additions of alum and cationic polymer and flocculation time were found to be very important. This is discussed in detail in Section 6.5.4.

- Backwash method: Any conventional washing procedure can be used. The usual backwashing rate is 37.5–60 $m^3/m^2 \cdot h$. In the case of dual-media and multi-media filters, the backwashing rate should be chosen in such a way as to minimize the intermixing of media. Details are given elsewhere (Baumann, 1978; Kawamura, 1975). The total backwash water volume can be reduced by a combined air-water backwash method. Direct filtration needs frequent backwashing. When a plant contains more than four direct filtration units, interfilter backwashing can be used. Cleasby (1993) recommends an inlet restriction as the method of flow control when interfilter backwashing is used.

- Filter-run length: Filter-run length depends on the raw water characteristics, the coagulant dose, the mixing energy input for floc formation (extent of pretreatment), the media size, the filtration rate, etc. As a guideline for design, the design parameters should be chosen in such a way that the filter run is at least 12 hours.

- Head requirement: Since the filtration rate declines with time in a direct filtration plant, the total head requirement for a direct filtration plant is less than that for a constant-rate filtration plant with the same flow rate. The head previously consumed in underdrains and piping during the early high-rate part of the filter cycle decreases with the square of the filtration rate, and becomes available to overcome the clogging headloss late in the cycle. Also, the headloss due to dirt accumulation is reduced as the rate declines. The total head requirement is typically about two-thirds of that required for a constant-rate plant (Cleasby, 1993).

- Use of a standby filter: United States regulations often require the use of a standby filter in a bank of declining rate filters. This one filter remains off-line after it is backwashed and is brought on-line as the next dirty filter needs to go off-line for backwashing. This arrangement has two other advantages beside avoiding the spikes in level and filtrate turbidity. First, the filter that goes off-line for backwashing can be allowed to continue to filter after the inlet flow is stopped. It will slowly drop in level as it filters to the clear well. This may even take more than an hour in some cases. By doing this, the pretreated water above the filter is not wasted in the desire to finish the backwash operation in a hurry. Also, due to the reduction in spike, the filter box depth can be reduced. The clean filter will come on-line with a lower total head and therefore will start at a lower filtration rate. Such use of a standby filter means that the extra filter is not truly redundant (standby) as intended by the regulations (Cleasby, 1993).

6.5.2 Economics

Cost data for direct filtration in the United States have been reported by Culp (1977); Logsdon et al. (1980); McCormick and King (1980); Monscvitz et al. (1978); Mueller and Conley (1981); Tate and Trussel (1980); and Tate et al. (1977). Culp (1977) states that the capital cost saving in direct filtration could be as high as 30%, and that a saving of 10–30% in chemical cost could be achieved. The costs for polymer may be higher than in conventional plants, but these costs are more than offset by the lower costs for the coagulant. Monscvitz et al. (1978) report that to add sedimentation units to one water treatment plant in Las Vegas would require more than 50% additional capital expenditure. Tate et al. (1977) also report capital cost savings of approximately 30% for the Utah Valley water treatment plant. When the 272,500 m^3/d Toronto water treatment plant was doubled in capacity by adopting direct filtration, the cost saving was \$4.8 million (or 35%) as compared with conventional treatment (Tredgett, 1974). Similarly, studies in Virginia, indicate that the use of direct filtration for waters of turbidity less than 10 NTU should result in savings of 10–30% in total annual costs (McCormick and King, 1980, 1982).

It should be noted that the capital savings from omission of settling basins can be slightly offset by the reduction in the length of the filter runs (Logsdon, 1978). Cost comparisons of conventional and direct filtration plants should be made on the basis of designs that permit optimum economy for each mode, rather than assuming fixed filter-run lengths for both modes. Figures 6.13 and 6.14 show the estimated curves for total capital costs and operation and maintenance costs, respectively. A cost comparison made by Mueller and Conley (1981) indicated that the smaller the plant, the more the direct filtration process should be given preference over the conventional system. For instance, at a capacity of 200 m^3/d, it is twice as cheap. In each instance, the capital cost includes an allowance for the

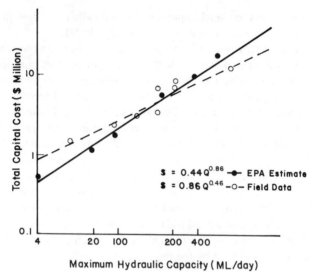

FIGURE 6.13   Total capital cost of direct filtration. (Logsdon et al., 1980)

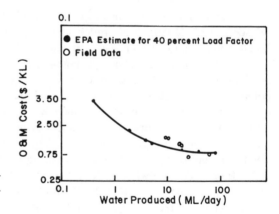

FIGURE 6.14   Operation and maintenance costs of direct filtration. (Logsdon et al., 1980)

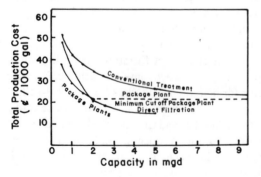

FIGURE 6.15   Comparison of package plants, direct filtration, and conventional treatment. (Clark and Morand, 1981)

equipment, its installation, and its foundations and housing. Intake structures, filtered water storage, high-service pumps, and distribution systems were excluded in this analysis.

Clark and Morand (1981) reached a similar conclusion when they compared the three treatment options available for small water supply systems, namely (1) conventional treatment (flocculation, sedimentation, filtration), (2) direct filtration, and (3) package treatment plants. Costs for annual operation and maintenance and capital costs over a range of specific flow levels were calculated, and cost equations developed. Total production costs for the three treatment alternatives are plotted versus plant capacity in Figure 6.15, which assumes that the average flow is 70% of the designed or capacity flow. All of the costs have been standardized with 1979 as a base year. It can be seen that in the production range below 2 mgd (7600 m^3/d), package treatment represents a lower cost alternative than do conventional or direct filtration. With plant capacities over 7600 m^3/d, direct filtration is significantly cheaper than the other two methods.

6.5.3 Advantages and Limitations

Advantages

- Direct filtration processes are normally found to be efficient and cost-effective for raw waters of relatively high quality (with turbidity less than 25 NTU).
- Direct filtration requires only that the colloids in raw water be destabilized into a small filterable floc. It is unnecessary to produce a settleable floc, which is more difficult to filter and far more expensive both in terms of chemicals and plant operation. Thus, a shorter flocculation time is required to form small flocs, and this reduces the power cost.
- Consequently, there is a substantial reduction in chemical dosages, of about 20–30%. This also results in decreased sludge production and thus less maintenance.
- The key benefit of direct filtration is cost savings. The omission of large sedimentation basins results in lower plant construction costs, and possible savings on land cost.
- Likewise, a reduction in operational and maintenance costs is obtained because less equipment is involved.
- The operation and maintenance of a direct filtration plant is simpler and easier, as compared with operating and maintaining a conventional plant.

Limitations

- Due to the elimination of the sedimentation process, the backwashing of the filter becomes more frequent. Also, since all the impurities are removed in the filter, more suspended solids are retained in the pores of the filter media, which requires a large amount of backwash water. Some of the experiences indicate that the backwash water used in direct filtration is as high as 6% of the water volume produced (Culp, 1977). Therefore, this shortcoming has to be taken into cost calculations before selecting an appropriate filtration method.

• Due to the shorter retention time between the application of coagulants and filtration and the greater loading applied to the filter, a significant amount of contaminated water enters the distribution system before the problem is discovered.

• For the same reason, more operator's vigilance is required. The chance of operational error is also greater than with the conventional treatment method. In order to mitigate this effect, continuous monitoring of effluent turbidity at each filter is a must (Logsdon, 1978).

• In the treatment of raw water containing a high concentration of coliform organisms, the bacteriological quality of the product water may not satisfy the public health requirements.

6.5.4 Application Status

Canada

As early as 1964, direct filtration was used in Toronto, when an existing plant with the maximum capacity of 72 mgd (272,520 m³/d) on Lake Ontario was converted to direct filtration (Hutchison and Foley, 1974; Tredgett, 1974). The use of alum plus polyelectrolyte when needed, followed by filtration through a dual-media filter with 18" (46 cm) of coal and 12" (30 cm) of sand produced high-quality effluent (less than 0.3 NTU). There was little change in the effluent turbidity for the filtration rates of 2.4–7.2 gpm/sq ft (5.9–17.6 m³/m²·h). Diatoms in the raw water had a marked influence on the length of the filter runs, but this problem could be overcome by using a coarse medium (such as coal) in the dual-media filters. As of 1976, seven direct-filtration plants existed in Ontario, and plans were underway for the construction of up to six additional plants to serve localities on Lake Ontario, Lake Huron, and Lake Superior (Hutchison, 1976).

United States

Diatomaceous earth and granular media direct filtration has been used at a number of full-scale plants with capacities from less than 1 mgd (3780 m³/d) to above 100 mgd (378,000 m³/d). Generally, diatomaceous earth plants are smaller, on the order of 1 to 10 mgd (3780–37,800 m³/d; Logsdon, 1978). Biggest plants include a 200-mgd (757,000 m³/d) plant at Las Vegas, Nevada, constructed in 1971, a 60-mgd (227,100 m³/d) plant for Springfield, Massachusetts, and a 30-mgd (113,550 m³/d) plant at Duluth, Minnesota, completed in 1976. After several years of operation, the plant at Duluth demonstrated that the process can be an effective and efficient means of providing high-quality treated water when proper design and operation parameters are adhered to (Hagar and Elder, 1981).

At the Springfield plant, raw water is conducted to the headworks of the conditioning basins. A channel running down the center of the conditioning structure feeds 14 sets of basins, 7 sets on each side. Each set of basins consists of two rapid-mix chambers in series, followed by two slow-mix chambers. A detention time of 30 minutes, believed conservative, is provided in the conditioning basins. Alum or iron salts may be used as the prime coagulants. The filter media consist of 24" (61 cm) of 1.0–1.1 mm effective size anthracite coal and 12" (30 cm) of 0.45 mm effective size silica sand. It also operates at a filtration rate of 12.5 $m^3/m^2 \cdot h$ (Sweeney and Prendiville, 1974). The plant at Las Vegas operates at a filtration rate of 12.2 $m^3/m^2 \cdot h$ and the filter media consist of 20" (51 cm) of 0.60–0.70 mm effective size anthracite and 10" (25 cm) effective size silica sand (Spink and Monscvitz, 1974; Monscvitz et al., 1978). Odor problems occur in two distinct periods — the spring and the fall — due to algal blooms. Activated carbon is added in the mixing chamber immediately ahead of the filters to remedy this problem (Spink and Monscvitz, 1974).

Australia

A detailed pilot-scale study was conducted with good-quality Warragamba dam water, Sydney, Australia, at a filtration rate of 13 m^3/m^2h with 2-m-deep 1.7 mm coal media. The results indicated that the unit filter run volume (UFRV) increased from 265 m^3/m^2 at an alum dose of 28 mg/L to about 540 m^3/m^2 with 6 mg/L alum and 1 mg/L cationic polymer (Cat Floc CL). Here UFRV is a measure of water production per unit filter area (units m^3/m^2) (UFRV = filtration time × filter run time) (Craig et al., 1993). Filter run time in these experiments was around 16 hours with alum alone and up to 30 hours with alum in combination with cationic polymer. A nonionic polymer filter aid was found to play an important role in maximizing filter performance. Detention time between the addition of the primary coagulant and cationic polymer and the addition of cationic polymer and filter aid was found to be very important, particularly at low temperatures. A lag time of at least 20 seconds between alum and cationic polymer addition was required under cold water conditions, while a lag time of 100 to 200 seconds between cationic and nonionic polymer addition was required to achieve maximum filter run time (Murray and Roddy, 1993b). A good flash mixing was found to give the best results for low alum dose coagulation. High-energy, short-time flash mixing can be achieved by injection mixers.

In the same study, preoxidation using chlorine or ozone improved the direct filtration performance significantly. The UFRV increased from 540 m^3/m^2 to 670 m^3/m^2 and 1200 m^3/m^2, respectively, when chlorine and ozone were used as oxidant prior to coagulation. Precoagulation oxidation allowed further reductions in alum dose by commencing microflocculation and by breaking down the organic molecules that contribute to color in the water. The ozone dose used was 5 mg/L. However, ozone was not found to be effective for high turbid waters.

Filter media selection is very important in direct filtration. A detailed study conducted with Warragamba dam water showed that coarse dual media produced a 30% increase in UFRV compared with 2.0 mm sand and filter coal. Narrowly graded coal (uniformity coefficient (UC) of 1.28) produced a 10% increase in UFRV compared with a coal with a higher UC of 1.4.

West Africa

Direct filtration can offer an economic advantage in West Africa, owing to the economy of a low alum dosage. The most striking examples are in Bamako, Mali, and Kano, Nigeria, (Wagner and Hudson, 1982). Both pay a high price for acquiring alum and hauling it to the treatment plant. The price of alum in Kano is over US $400/ton, and in Bamako it is over U.S. $700/ton.

The alum dose at the time of testing at Kano was 20 mg/L to treat water of turbidities in the range of 20–24 NTU. The water at Kano has turbidities of 30–40 NTU at its peak, and is clear during most part of the year. The average alum dose during the wet season is 26 mg/L, whereas in the dry season it is 15 mg/L.

Both of these water sources have been shown to be good candidates for direct filtration, and both cities are proceeding toward pilot filter testing. The effluent produced by the bench-scale work was well below the World Health Organization turbidity limit of 5 NTU.

EXAMPLE 6.1

Design a rapid sand filter to treat 2.27 ML/day supply. The filter works 12 hours per day. Assume the rate of filtration as 1123 ML/d·ha (1 ha = 10,000 m^2 = 10^4 m^2).

Solution

Quantity of water to be treated per hour = 2.27/12 ML/h
Filtration rate = 1123/24 ML/h/ha
Filtering area required = (2.27/12) / (1123/24) ha
$$= 0.00404 \text{ ha}$$
$$= 40.4 \text{ m}^2$$

If two units are provided, each unit has an area of 40.4/2 = 20.2 m^2.

Assume washing time of 20 minutes, which is nearly 3% of 12 hours working. Therefore, adding 3% more to the area of 20.2 m^2:

Area of each unit = 20.2 + 3/100 × 20.2
$$= 20.2 + 0.606$$
$$= 20.806 \text{ m}^2$$
Each filter = 6.0 m × 3.5 m size

EXAMPLE 6.2

A slow sand filter is to be designed for a design flow of 304 m³/d. Calculate the area of the filter functioning at the different modes of operation as follows:

1. 24 hours of constant-rate (CR) filtration
2. 16 hours of CR and 8 hours of declining-rate (DR) filtration.
3. 8 hours of CR and 16 hours of DR filtration.
4. 4 hours of CR and 8 hours of DR filtration followed by similar CR and DR filtration operations.

Assume a filtration velocity of 0.1 m³/m²·h.

Equation for the filter area (A) calculation is as follows (IRC, 1981):

$$A = Q/(va + b)$$

where v = filtration velocity (m³/m²·h), Q = design flow (m³/d), a = number of hours of constant-rate filter operation/day, b = constant, = 0.5 (for a declining-rate operation of 8 hours), = 0.7 (for a declining-rate operation of 16 hours).

Solution

1. A = (304 m³/d)/ (0.1 × 24 m/d) = 126.7 m²
2. A = (304 m³/d)/ (0.1 m/h × 16 h/d + 0.5) = 144.8 m²
3. A = (304 m³/d)/ (0.1 m/h × 8 h/d + 0.7) = 202.7 m²
4. A = (304 m³/d)/ 2(0.1 m/h × 4 h/d + 0.5) = 168.9 m²

A rule of thumb (IRC, 1981):

Q (m³/d)	Mode of operation
< 300	8 h of CR + 16 h of DR
300 < Q < 600	16 h of CR + 8 h of DR
> 600	24 h of CR

EXAMPLE 6.3

Design an underdrain (manifold-lateral) system of a slow sand filter with the following data:

- Design flow = 7.2 m³/h
- Filtration velocity = 0.1 m/h
- Headloss through the filter sand (of effective size of 0.25 mm and uniformity coefficient of 3) of 1 m depth is 6.6 cm at a filtration velocity of 0.1 m/h
- Headloss through the supporting gravel layers is 0.7 mm at this filtration velocity

Solution

Area of filter = 7.2 / 0.1 = 72 m²
Number of filters (N) = $Q^{0.25}$ (where Q in m³/h)
= $(7.2)^{0.25}$ = 1.64 ≈ 2

Assuming two filters in parallel, area of each filter = 36 m².

For optimum construction, width (W) and length (L) of the filter are related by the following equation:

$$W = L (N+1)/2N$$

where N = number of filters, = 3/4 L for two filters in parallel.

Area of each filter = L × W
= L × 3 L/4
= 36 m²

Therefore, select L = 7 m; W = 5.25 m.

Assuming a width of 6 m and following the lateral and manifold configuration shown in Figure 6.16, the length of each lateral will be 3 m. If the spacing between the laterals is 1.5 m, total number of laterals will be 10.

Flow rate through each lateral = Q / number of laterals
= 0.002/10
= 0.0002 m³/s

Here, a maximum value of Q is assumed (i.e., the entire flow is through one filter during the backwash of the other filter).

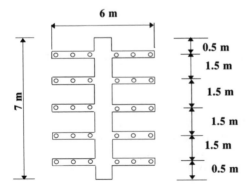

FIGURE 6.16
Lateral and manifold configuration.

Using 50 mm diameter PVC pipes as lateral pipes and a 15 × 15 cm concrete channel as the main pipe, velocity through each lateral (v_L) is given by

$$v_L = 0.0002 / (\pi/4)(50 \times 10^{-3})^2$$

$$= 0.10 \ m/s$$

(Since the velocity is less than 0.3 m/s, selected size of the lateral pipe is suitable.)

Headloss through laterals (H_L) = ($L_1 S_1$)/3 + v_L^2/2 g, where L_1 = length of lateral pipe, S_1 = hydraulic gradient = $10.2 \times Q_1^2 \ n^2/d^{5.33}$, Q_1 = flow rate through laterals (m^3/s), n = roughness coefficient (= 0.01 for small PVC pipes), d = diameter of lateral (m).

Substituting the values, S_1 = 0.000352. Therefore

$$H_L = (3 \times 000352/3) + 0.1^2/9.81$$
$$= 1.02 \ mm$$

Similarly, the headloss in the main pipe (H_m) can be calculated using the following equation:

$$H_m = (L_m S_m)/3 + v_m^2/2 \ g$$

where L_m = length of main pipe = 7 m, S_m = hydraulic gradient = $v_m^2 \ (n/R^{2/3})^2$, v_m = velocity in the main pipe, R = hydraulic radius, n = roughness coefficient.

Substituting the values, H_m value is found as 1.07 mm. Thus the total headloss through the underdrain system and gravel supporting layer = 1.02 + 1.07 + 0.7 = 2.79 mm. To have an even flow through the entire underdrain system, headloss through the underdrain system and gravel layers should be less than 1/4 of the headloss through the filter sand, i.e., 2.79 mm < 1/4 × 66 mm. Therefore, our assumptions on lateral-manifold dimensions are acceptable.

EXAMPLE 6.4

A single medium filtration system operates in a water works with sand as a filter medium of 100 cm depth. The effective size (d_{10}) and the uniformity coefficient (UC) of the sand are 0.5 mm and 1.5, respectively. The filtration rate is 7 $m^3/m^2 \cdot h$. It is required to increase the filtration rate to 10 $m^3/m^2 \cdot h$ by converting the filter into a dual-media filter by replacing a part of the sand medium with anthracite coal. It is required to maintain the total height of the media the same as the original depth (i.e., 100 cm) and the clean bed headloss the same as before. Calculate the anthracite coal size that can be used (Vigneswaran and Dharmappa, 1992). (See Figures 6.17 and 6.18.)

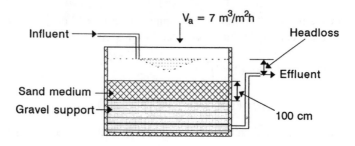

FIGURE 6.17   Before modification.

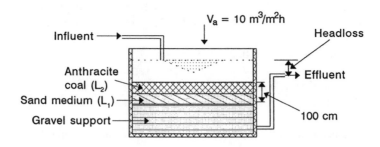

FIGURE 6.18   After modification.

Solution

For sand,

$(d_{10})_s = 0.5$ mm
UC = 1.5
$L = 100$ cm $= L_1 + L_2$
$v_a = 7$ m³/m²·h

Determination of the Suitable Grain Size of Anthracite Coal

Minimum fluidization velocity (v_{mf}) can be calculated from the following equation (Amirtharajah, 1978):

$$v_{mf} = 0.00381 \, (d_{60})_s^{1.82} \, [\rho \, (\rho_s - \rho)]^{0.94} \, / \, \mu^{0.88}$$

where $d_{60} = 60\%$ media size; ρ and ρ_s = densities of water and sand, respectively; μ = viscosity.

To minimize the intermixing of two media, fluidization velocity for sand $(V_{mf})_S$ = fluidization velocity for anthracite $(V_{mf})_A$.

$$\left\{ 0.00381 \, (d_{60})_s^{1.82} \left[\rho \, (\rho_s - \rho) \right]^{0.94} \Big/ \mu^{0.88} \right\} = \left\{ 0.00381 \, (d_{60})_A^{1.82} \left[\rho \, (\rho_A - \rho) \right]^{0.94} \Big/ \mu^{0.88} \right\}$$

$$(d_{60})_s / (d_{60})_A = (\rho_A - \rho / \rho_s - \rho)^{0.52} \quad = ((S_A - 1)/(S_S - 1))^{0.52}$$

Specific gravity of sand (S_s) and anthracite (S_A) are 2.65 and 1.50, respectively. Therefore,

$$(d_{60})_s / (d_{60})_A = (1.5 - 1 / 2.65 - 1)^{0.52} = 0.537$$

For sand,

$$d_{60}/d_{10} = UC = 1.5$$

$$(d_{60})_s = 1.5 \times 0.5 = 0.75 \text{ mm}$$

Therefore,

$$(d_{60})_A = 0.75/0.537 = 1.4 \text{ mm}$$

Uniformity coefficient for anthracite = 1.4.

Effective size of anthracite = $(d_{10})_A = (d_{60})_A/(d_{10})_A = 1.4/1.4 = 1$ mm.

Depth of the Filter Media Before Modification

Headloss in the existing filter (h_f) can be calculated from Kozeny's equation,

$$h_f/L = kv_a \left\{ \Upsilon(1 - f_s)^2 / gf_s^3 \right\} \left\{ 6/\Psi_s(d_{60})_s \right\}^2$$

$$= \left\{ 5 \times 0.194 \times 1.002 \times 10^{-2}(1 - 0.4)^2 /981 \times 0.4^3 \right\} [6/0.8 \times 0.075]^2$$

where v_a = filtration rate = 700/3600 = 0.194 cm/s, k = Kozeny's constant = 5, Υ = kinematic viscosity = 1.002×10^{-2} cm^2/s, Ψ_s = sphericity = 0.8, f_s = porosity = 0.4, h_f/L = 0.558. Headloss in the existing filter = $0.558 \times 100 = 55.8$ cm.

After Modification by Placing Anthracite Coal

$$h_f = 55.7 = (kv_a \tau/g) [\{(1 - f_s)^2/f_s^3\} \{\Psi_s(d_{60})_s\}^2 L_1 + \{(1 - f_A)^2/f_A^3\} \{6/\Psi_A(d_{60})_A\}^2 L_2]$$

Substituting the values, f_A = 0.5; Ψ_A = 0.7; and v_a = 10 m/h = 0.278 cm/s, we get

$$55.8 = 0.799 L_1 + 0.1065 L^2$$

We have,

$$L_1 + L_2 = 100; L_1 = 100 - L_2$$

$$L_2 = 35 \text{ cm}$$

$$L_1 = 100 - 35 = 65 \text{ cm}$$

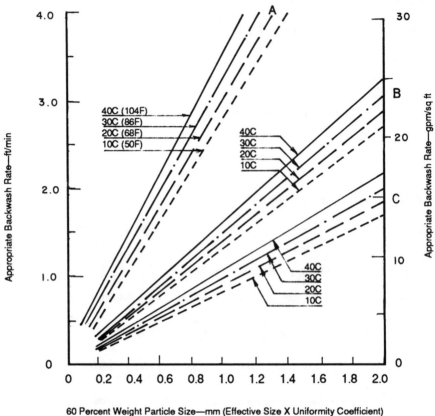

A—Filter sand (sg = 2.60); B—anthracite coal (sg = 1.50); C—spherical resin grains* (sg = 1.27)

FIGURE 6.19 Appropriate filter backwash rate for filter media. (Kawamura, 1975)

Backwashing Rate

Using the graph in Figure 6.19 with the following conditions:

$(d_{60})_s = 0.75$ mm
$(d_{60})_A = 1.4$ mm
Temperature = 20°C

Backwash velocity (v_b) can be calculated as 2.5 m/min.

EXAMPLE 6.5

The following data are available for a sand filter medium with an average size 0.7 mm and a depth of 1 m. Calculate the headloss through the stratified sand bed. Assume the filter velocity as 7.2 m/h, porosity of sand as 0.4, and sphericity coefficient (Ψ) as 0.8.

Mesh size (mm)	Percent fraction (by weight)
3.360–2.380	3.2
2.380–1.680	5.3
1.680–1.190	17.1
1.190–0.840	14.6
0.840–0.590	20.4
0.590–0.420	17.6
0.420–0.297	11.9
0.297–0.210	5.9
0.210–0.149	3.1
0.149–0.105	0.9

Kozeny's equation for clean filter bed headloss is:

$$\Delta H/L = (h_k \ v \ \Upsilon/g) \ (6/\Psi)^2 \ \{(1 - f)^2/f^3\} \ a_g^2 \ \Sigma \ p_i/d_i^2 = 483.45$$

where h_k = Kozeny's constant = 5, v = filtration rate = 7.2 m/h (or 0.2 cm/s), f = porosity = 0.4, Υ = kinematic viscosity = 0.01 cm²/s, g = acceleration due to gravity = 981 cm/s², a_g = specific surface of filter medium, Ψ = sphericity coefficient = 0.8. Specific surface of filter medium for stratified medium is given as:

$$a_g = \sum_{i=1}^{n} p_i/d_i^2$$

where p_i = weight fraction of filter medium of size d_i.

Mesh size (mm)	Percent fraction	Geometric mean size (mm)	p_i/d_i^2 (cm⁻²)
3.360–2.380	3.2	2.83	0.40
2.380–1.680	5.3	2.00	1.33
1.680–1.190	17.1	1.41	8.55
1.190–0.840	14.6	1.00	14.60
0.840–0.590	20.4	0.70	41.16
0.590–0.420	17.6	0.50	70.97
0.420–0.297	11.9	0.35	95.50
0.297–0.210	5.9	0.25	94.40
0.210–0.149	3.1	0.18	98.95
0.149–0.105	0.9	0.13	57.60

Headloss for Stratified Bed

$$\Delta H/L = h_k \left(v\gamma/g \right) \left\{ (1-f)^2/f^3 \right\} \left(6/\Psi \right)^2 \Sigma p_i/d_i^2$$

$$= 5 \left(0.2 * 0.01/981 \right) \left(0.6^2/0.4^3 \right) \left(6/0.8 \right)^2 \times 483.45$$

$$= 1.559$$

$$\Delta H = 1.559 * 1 = 1.559 \text{ m}$$

REFERENCES

Adin, A., Baumann, E. R. and Cleasby, J. L., The application of filtration theory to pilot-plant design, *J. AWWA*, 71(1), 17, 1979.

Amirtharajah, A., Optimum backwashing of sand filters, *J. Environ. Eng. Div. ASCE*, 104(EE5), 917, 1977.

Amirtharajah, A., Design of granular-media filter units, *Water Treatment Plant Design*, R. L. Sanks (Ed.), Ann Arbor Science, Michigan, 1978.

Anon., SWS unit, *World Water*, 24, May 1983.

AWWA Filtration Committee, The status of direct filtration, *J. AWWA*, 72(7), 405, 1980.

Baumann, E. R., Granular-media deep-bed filtration, *Water Treatment Plant Design*, R. L. Sanks (Ed.), Ann Arbor Science, Michigan, 1978.

Ben Aim, R., Sahnoun, A., Chemin, C., Hahn, L., Visvanathan, C. and Vigneswaran, S., New filtration media and their use for water treatment, *Proceedings, World Filtration Congress*, Nagoya, Japan, 1993.

Camp, T. R., Graber S. D. and Conklin, G. F., Backwashing of granular water filters, *J. Environ. Eng. Div.-Proc. ASCE*, 97(EE6), 903, 1971.

Cansdale, G. S., *Low-Cost Filtration System*, Report No. SCS/79/WP/80, South China Sea Fisheries Development and Coordinating Program, Manila, 1979a.

Cansdale, G. S., *Report on Second Regional Consultancy Low-Cost Filtration System*, Report No. SCS/79/WP/84, South China Sea Fisheries Development and Coordinating Program, Manila, 1979b.

Clark, R. M. and Morand, J. M., Cost of small water supply treatment systems, *J. Environ. Eng. Div. ASCE*, 107(EE5), 1051, 1981.

Cleasby, J. L., Declining rate filtration, *Water Science and Technology*, 27(7–8), 11, 1993.

Conley, W. R. and Hsiung, K., Design and application of multimedia filters, *J. AWWA*, 61(2), 97, 1969.

Craig, K., Naylor, R., Murray, B. and Roddy, S., The use of alum versus ferric salts with cationic polymer in contact filtration studies, *Proceedings of the 15th AWWA Federal Convention, Gold Coast, 1059-1064*, April 1993.

Culp, R. L., Direct filtration, *AWWA J.*, 69, 375, 1977(7).

Fair, G. M., Geyer, J. C. and Okun, D. A., *Water and Wastewater Engineering*, Vol. 2, John Wiley & Sons, New York, 1968.

Frankel, R. J., Operation of the coconut fiber/burnt rice husks filter for supplying drinking water to rural communities in Southeast Asia, *Am. J. Public Health*, 69(1), 75–76, 1979.

Gadkari, S. K., Raman, V. and Gadkari, A. S., Studies of direct filtration of raw water, *Indian J. Environ. Health*, 22(1), 57–65, 1980.

Hagar, C. B. and Elder, D. B., Filtration processes, *Proc. AWWA 1981 Annual Conference*, American Water Works Association, 1981.

Hamann, C. L. and McKinny, R. E., Upflow Filtration Process, *J. AWWA*, 60(9), 1023–1039, 1968.

Huisman, L., Developments in village scale slow sand filtration, *Progress in Water Technology*, 11, 159–165, 1978.

Hutchison, W. R., High-rate direct filtration, *J. AWWA*, 66(6), 292–298, 1976.

Hutchison, W. and Foley, P. D., Operational and experimental results of direct filtration, *J. AWWA*, 66(1), 79–87, 1974.

IRC—International Reference Center for Community Water Supply and Sanitation, *Slow Sand Filtration for Community Water Supply in Developing Countries*, Bulletin Series No. 16, The Netherlands, 1981.

Jahn, S. A. A., *Traditional Water Purification in Tropical Developing Countries—Existing Methods and Potential Application*, German Agency for Technical Cooperation (GTZ), Eschborn, Federal Republic of Germany, 1981.

Kawamura, S., Design and operation of high-rate filters—Part I, *J. AWWA*, 67(10), 535, 1975.

Kawamura, S., *Integrated Design of Water Treatment Facilities*, Wiley Interscience, 1991.

King, P. H. and Amy, W. T., *The Potential of Direct Filtration for Water Treatment in Underdeveloped Countries*, unpublished manuscript, 1979.

Landis, D. M., De laval immedium up-flow for steel rolling mill waste water, *Eriez Magnetics Third Annual Executive Forum*, Erie, Pennsylvania, September, 1966.

Logsdon, G. S., Direct filtration—past, present, future, *Civ. Eng.-ASCE*, 70(7), 68, 1978.

Logsdon, G. S., Clark, R. M. and Tate, C. H., Direct filtration treatment plants: Costs and capabilities, *J. AWWA*, 72(3), 134, 1980.

Mazumdar, B., *A Study of the Phenomenon of Intermixing in Dual-Media Filters and its Effects on Filter Performance*, AIT Research Study, Asian Institute of Technology, Bangkok, 1984.

McCabe, W. L. and Smith, J. C., *Unit Operations of Chemical Engineering*, 2nd ed., McGraw-Hill, New York, 1967.

McCormick, R. F. and King, P. H., *Direct Filtration of Virginia Surface Waters: Feasibility and Costs*, Bulletin No. 129, Virginia Polytechnic Institute and State University, Blacksburg, 1980.

McCormick, R. F. and King, P. H., Factors that affect use of direct filtration in treating surface waters, *J. AWWA*, 74(5), 234, 1982.

Monscvitz, J. T., Rexing, D. J., Williams, R. G. and Heckler, J., Some practical experience in direct filtration, *J. AWWA*, 70(10), 584, 1978.

Murray, B. and Roddy, S., Precoagulation oxidation and high rate contact filtration, *Proceedings of the 15th AWWA Federal Convention*, April 1993, Gold Coast, 1093, 1993a.

Murray, B. and Roddy, S., Treatment for Sydney's water supplies—An overview of pilot and prototype plant studies, *AWWA Water J.*, 20(1), 17, 1993b.

Nigam, J. P., Low cost filtration plants for water supply of Ardha-Kumbha Mela, 1980 at Hardwar, *J. Inst. Eng. (India)—Environ. Eng. Div.*, 61(3), 117, 1981.

NWS & DB, *Design Manual D3,* National Water Supply and Drainage Board, Sri Lanka, 1988.

Okun, D. A., Digest of sanitation engineering research reports, upflow filtration, *Public Works*, 98(6), 194–201, 1967.

Paramasivam, R., Mhaisalkar, V. A. and Berthouex, P. M., Slow sand filter design and construction in developing countries, *J. AWWA*, 73(4), 178, 1981.

Schulz, C. R. and Okun, D. A., Treating surface waters for communities in developing countries, *J. AWWA*, 75(5), 524, 1983.

Seelaus, T. J., Hendricks, D. W. and Janonis, B., Design and operation of a slow sand filter, *J. AWWA*, 78(12), 35, 1986.

Spink, C. M. and Monscvitz, J. T., Design and operation of a 200-mgd direct-filtration facility, *J. AWWA*, 66(2), 127, 1974.

Sweeney, G. E. and Prendiville, P. W., Direct filtration: An economic answer to a city's water needs, *J. AWWA*, 66(2), 65, 1974.

Tate, C. H. and Trussel, R. R., Recent developments in direct filtration, *J. AWWA*, 72(3), 165, 1980.

Tate, C. H., Lang, J. S. and Hutchinson, H. L., Pilot plant tests of direct filtration, *J. AWWA*, 69(7), 379, 1977.

Thanh, N. C. and Hettiaratchi, J. P. A., *Surface Water Filtration for Rural Areas—Guidelines for Design, Construction, and Maintenance,* Environmental Sanitation Information Center, Bangkok, 1982.

Tredgett, R. G., Direct filtration studies for Metropolitan Toronto, *J. AWWA*, 66(2), 103, 1974.

Valencia, J. A., Some basic ideas on establishing a water treatment technology adapted to developing countries, *Proc. Intl. Train. Sem. on Community Water Supply in Developing Countries*, Bulletin Series No. 10, International Reference Center for Community Water Supply and Sanitation, The Netherlands, 1977.

Vigneswaran, S., Tam, D. H. and Visvanathan, C., Water filtration technologies for developing countries, *Environmental Sanitation Reviews*, ENSIC, AIT, Bangkok, 12, 1983.

Vigneswaran, S. and Dharmappa H. B., Unpublished Notes on Wastes and Wastewater Treatment, University of Technology, Sydney, 1993.

Vigneswaran, S. and Ngo, H. H., Trends in water treatment technologies, *Proceedings of Korea — Australia Joint Seminar on the Recent Trends in Technology Development for Water Quality Conservation*, Seoul, Korea, June 1993, 145–155, 1993.

Vigneswaran, S. and Ben Aim, R., Improvements on rapid filter, *Water Wastewater and Sludge Filtration*, CRC Press, Boca Raton, Florida, 1989.

Visscher, J. T., Slow sand filtration: Design, operation and maintenance, *J. AWWA*, 82(6), 67, 1990.

Wagner, E. G. and Hudson, H. E., Low-dosage high-rate direct filtration, *J. AWWA*, 74(5), 256, 1982.

Werellagama, D. R. I. B., *Application of the Floto Filter Unit in Contact Flocculation Filtration of Surface Waters*, Masters Thesis EV 93-26, Asian Institute of Technology, Bangkok, 1993.

7: Water Treatment for Specific Impurities Removal

CONTENTS

7.1 IRON AND MANGANESE REMOVAL

7.1.1 Introduction

The presence of both iron and manganese above a particular concentration is unde-sirable in water, and excess should be removed. Though there are no harmful effects for humans from drinking waters containing iron and manganese, they are unacceptable from an aesthetic point of view. In fact iron is an essential element required for humans for the formation of hemoglobin, which is essential in transporting oxygen from the lungs to tissue cells. For proper nutrition, human adults require 10 to 20 mg/L of iron intake per day, and deficiency of iron in the human body causes anemia (Zainuddin, 1985). The problems of Fe and Mn usually occur when the raw water source is groundwater, but they can also be found in anaerobic surface waters and where industrial wastes are discharged. Marked high levels of iron and manganese are found in the anoxic hypolimnion of lakes and impoundments under eutrophication (Mouchet, 1992).

Fe and Mn are natural constituents of soils and rocks. They are normally present in highly insoluble forms, but are brought into solution by anaerobic conditions or by the presence of CO_2 (as in a deep well or the bottom of a reservoir). Nuisance conditions can occur when the following concentrations are exceeded:

- 0.3 mg/L iron
- 0.1 mg/L manganese

Iron exists in solution in the ferrous state, usually as ferrous bicarbonate. It can only remain in solution in the absence of oxygen, and generally when the pH is below 6.5. When such water is exposed to air, the soluble ferrous bicarbonate is oxidized to insoluble ferric hydroxide ($Fe(OH)_3$). Thus, for example, a deep well water sample, upon exposure to the atmosphere, may become opalescent and discolored due the oxidation of the soluble salt to the insoluble form, and will form a deposit. Even when iron is not present in the source water, it is usually acquired to some extent from contact with iron pipes and fittings.

Manganese occurs in a similar way to iron, and they may both be present at the same time. Deposits and discoloration due to manganese are black rather than rust-colored as for iron. The Mn deposits are of higher density and are less easily flushed out. Mn oxidizes less readily than iron and therefore is more difficult to remove.

7.1.2 Problems Posed by the Presence of Iron and Manganese

The following list gives an idea of the difficulties to consumers caused by the presence of iron and manganese in raw water.

1. Introduce objectionable bitter or metallic taste.
2. Introduce reddish-brown or black color to water.
3. Interfere with plumbing fixtures by leaving deposits.
4. Stain clothes during laundering.
5. Support growth of microorganisms such as *Clonothrix* and *Crenothrix*, particularly when organic matter is present. These bacteria may accentuate the deposition problems and may produce sulfides. This could clog pipelines and cause taste and odor problems.
6. Cause difficulty in ion-exchange softening unit by clogging and coating the exchange medium.

Iron and manganese may also accelerate biological growths in the distribution system, further exacerbating taste, odor, and color problems. To prevent these difficulties, various regulatory agencies have put forward standards to control iron and manganese concentrations. For example, the US EPA has specified iron and manganese levels to be within 0.3 and 0.05 mg/L, respectively, in drinking water.

Problems due to Fe and Mn in distribution mains may be minimized by:

- prior removal by appropriate treatment
- protecting iron/steel mains with bituminous linings, or using noncorrosive materials
- avoiding dead-end mains
- avoiding disturbances in the water flow
- flushing periodically

Organic substances found in surface waters and groundwaters can form soluble complexes with Fe and Mn. These complexed forms of iron and manganese may not be satisfactorily oxidized to an insoluble form during treatment and therefore will not be removed by solid-liquid separation processes. The use of strong oxidants may be required to treat such waters. In contrast, some organic compounds (e.g., citrate) are capable of catalyzing the rate of Fe(II) oxidation (AWWA Trace Organics Substances Committee, 1987).

7.1.3 Different Treatment Methods for Removal of Iron and Manganese

The treatment adopted in the removal of different impurities will have a bearing on the choice of method for iron removal. For example (USAID Sri Lanka Project, 1989):

- iron is usually accompanied by high amounts of free CO_2; removing only the iron and leaving the free CO_2 could cause corrosion of pipes and mains
- lime softening if required for hardness removal, will also be effective for removing the iron and CO_2
- if organics removal/bacterial disinfection is necessary, chlorination may be required, which will assist in the iron and manganese removal

In surface water, due to the presence of oxygen, the iron and manganese are found in oxidized form and can be removed relatively easily. Groundwater generally is deprived of oxygen and so has high content of reduced iron, which can be removed by aeration or chemical oxidation. Of the technologies available for iron and manganese removal, the following ones have found widespread application in water supply (Mouchet, 1992):

- Aeration followed by sand filtration (or dual-media filtration), often complimented by a contact tank, settling, or flotation and the addition of chemicals.

- Chemical oxidation (without preaeration) followed by filtration.
- Filtration with a special medium that acts as an ion or electron exchanger, e.g., manganese greensand, zeolites of volcanic origin (tectosilicates), sand that is naturally/artificially coated with manganese dioxide to simulate a natural greensand, or adsorption with activated carbon.
- Magnesium oxide and diatomite precoat filtration (analogous to manganese removal by magnesium hydroxide).
- Conventional treatment combined with lime softening.
- Sodium silicate, phosphates, or polyphosphates as sequestering agents.
- Biological treatment methods.
- *In situ* treatment, in which oxygenated water is introduced into the aquifer by means of feed wells, thus creating a treatment area around the main well. This method is based on the combined efficiency of simultaneously occurring physical, chemical, and biological phenomena (Viswanathan, 1989).

Among the above techniques, aeration-filtration is commonly used in developing countries. Chemical precipitation is more suitable for water with a higher concentration (Fe >5.0 mg/L). The potassium permanganate-manganese greensand method is suitable for low to moderate concentrations of iron and manganese, about 0–5 mg/L. The prechlorination-filtration process is generally recommended for low iron concentrations (e.g., less than 2.0 mg/L). The ion exchange method is used only for small quantities, while activated carbon adsorption is relatively expensive. Biological treatment involves a combination of both physicochemical and biological removal mechanisms. The biological method is extensively used in European countries such as Holland and Germany, and is advantageous primarily when the water contains iron, manganese, and ammonia at the same time.

Oxidation-Filtration Methods

Oxidation is a common preventive treatment for iron and manganese. The relatively soluble Fe(II) and Mn(II) are oxidized into insoluble Fe(III) and Mn(III, IV). Any organic complexing agents present in the raw water are also oxidized. Subsequently, the Fe(III) and Mn(III, IV) precipitates are removed by filtration (or sedimentation and filtration). Unoxidized Fe(II) and Mn(II) can be adsorbed by hydrous ferric and manganese coatings on the filter media, where the reduced forms are subsequently oxidized to Fe(III) and Mn(III, IV). The efficiency of oxidizing agents as regards to Fe and Mn removal is given in Table 7.1.

If only iron is present, simple aeration followed by filtration will be effective. Oxidation is rapid if pH is above 7.5. pH adjustment may be necessary to increase the removal efficiency. Aeration followed by sedimentation and rapid sand filtration is required when the amount of Fe is high or when Mn is present. Chemical addition is necessary for removal of high CO_2 levels, or if larger amounts of soluble components of iron and manganese are present.

Table 7.1 Relative Efficiency of Oxidizing Agents for Iron and Manganese Removal

Metal ion to be removed	Oxidizing agent				
	Air	**Cl₂**	**ClO₂**	**O₃**	**KMnO₄**
Fe	Very good	Very good	Very good	Excellent	Good
Mn	No effect	Good	Very good	Excellent	Excellent

Prechlorination

In iron removal, prechlorination is sometimes advisable as it accelerates the oxidation of iron and allows ammonia to be removed chemically. It also removes growths of iron bacteria. Prechlorination may be used for oxidation of Fe and Mn, but aeration is preferred. Prechlorination ahead of conventional treatment (coagulation, sedimentation, filtration) at pH values of 6.7–8.4 should provide adequate removal of iron and manganese in surface waters containing significant concentrations of organic material. It may also be harmful if the chlorination break-point cannot be reached, because in such cases prechlorination stops certain forms of biological iron removal and nitrification without exerting a sufficient oxidation power. Preliminary tests are therefore required before any prechlorination treatment is attempted (Degremont, 1991).

Iron Removal Without Clarification (Aeration-Filtration)

Aeration assists in dissipation of free carbon dioxide, and oxidation and precipitation of iron. Free CO_2 should be reduced to less than 10 mg/L to remove corrosive properties. Hydrogen sulfide, if present, should also be dissipated. This process is based on the principle of oxidation of iron by oxygen from air, using different types of aerators. This process is applied to raw water with a maximum iron concentration of 5 mg/L, and with no other unfavorable characteristics (manganese, color, turbidity, humic acids); a low level of ammonia may be tolerated. In some cases, aeration can be used for raw waters with a higher iron content (up to 10 mg/L). Oxidation is rapid with a pH over 7.5. pH adjustment may be necessary to increase the removal efficiency.

The first stage of iron removal is oxidation of Fe^{++} by oxygen from air. The aeration process may take place

1. at atmospheric pressure for installations operated by gravity
2. under pressure in which compressed air is blown into oxidation towers with contact materials

The aeration at atmospheric pressure often provides a cheap means of removing aggressive carbon dioxide, which otherwise requires expensive neutralization treatment. The advantage of the second type (compressed air) is that it can be operated at system delivery pressure without pumping. Both types are used in France (Mouchet, 1992), and in the pressurized iron removal plants, hard porous volcanic lava is used as the contact material, which ensures maximum contact of the water with the air. The filter is equipped for backwashing and air scour. This type of plant is well suited for small and medium capacity systems. Gravity units are generally reserved for waters with a high carbon dioxide or hydrogen sulfide content and for systems with high capacity.

The rate at which Fe^{++} is oxidized depends upon various factors, such as temperature, pH, dissolved oxygen, and iron content itself. The reaction can be expressed as follows:

$$4Fe^{2+} + O_2 + 8OH^- + 2H_2O \longrightarrow 4Fe(OH)_3 \qquad (7.1)$$

The above equation shows that 0.14 mg oxygen is required to oxidize 1 mg of iron. The oxidation time may be considerably affected by the catalytic action of

- existing deposits
- certain anions in the water (particularly silicates and phosphates)
- metallic catalysts introduced into the water during the treatment, such as traces of copper sulfate which may exert a strong influence over the oxidation of iron and manganese by oxygen or chemical oxidants

According to the method used, the precipitate formed may contain larger or smaller proportions of ferrous carbonate which, being more crystalline than ferric hydroxide, dictates the effective size of the filter media. The normal filter-medium size ranges from 0.5 to 1.7 mm and filtration rate from 5 to 20 $m^3/m^2 \cdot h$ or even higher. Filters with a homogenous layer of sand are suitable for the majority of cases, provided that the filtration rate, grain size, and depth of the bed are carefully designed. In certain problem cases, dual-media filters (anthracite and sand) or filters with sand artificially coated with manganese dioxide give good results.

Some substances such as humic acids, silicates, phosphates, or polyphosphates act as inhibitors of the precipitation and filtration of ferric hydrate. These effects can be controlled by additional treatment such as oxidation ($KMnO_4$, ozone), coagulation (aluminum sulfate), or flocculation (alginate), depending on the circumstances (Degremont, 1991).

Figure 7.1 shows a schematic diagram of an aeration filtration unit used for iron removal. Aeration-filtration is the simplest method for removal of iron and manganese; however, it has some major disadvantages, such as:

1. high initial cost
2. necessity of additional retention time when soluble manganese concentration exceeds 1 mg/L
3. necessity of supplementary chemical treatment to decrease manganese concentration below 1 mg/L

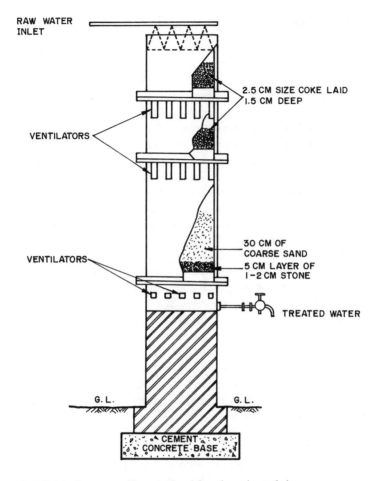

FIGURE 7.1   Iron removal by natural and forced aeration technique.

Iron Removal With Clarification

A clarification stage, as shown in Figure 7.2, needs to be inserted prior to a filtration unit under the following conditions:

- high level of iron in the raw water (> 5–10 mg/L), resulting in an undue amount of precipitate, which subsequently will clog the filters (Mouchet, 1992)
- presence of color, turbidity, humic acids, complex forming agents, etc., which require the addition of the coagulant at a dosage rate higher than 10 mg/L (Degremont, 1991)

Sludge-blanket-type settlers are especially well suited, because the catalytic effect of sludge contact improves the reaction kinetics. An aeration stage must precede clarification if the raw water is deprived of oxygen.

Small-Scale Iron Removal Plant for Rural Water Supply

Vigneswaran and Joshi (1990) presented a simple iron and manganese removal plant, suitable for household water supplies in rural areas where groundwater is extensively used for drinking and other domestic purposes. This unit (Figure 7.3) consists of a contact filter to oxidize iron and manganese and a matrix filter to remove the precipitates. The aeration (in part A) oxidizes soluble Fe and Mn into insoluble ferric and manganese oxide hydrates, which can be filtered out. Raw water is sprayed on the filter bed by means of a perforated PVC pipe. Ventilation holes supply air to promote oxidation. These holes also serve as outlets for backwash water as well as outflow water from the filter. The contact filter simply consists of coarse gravel or broken road stone (metal) in the size range 5–25 mm. It acts as a contact medium to enhance the oxygen transfer through aeration and turbulence.

The fixed-bed filter called the matrix filter (part B) consists of a coarse medium of 50 mm size, which is partly filled with sand. The matrix creates a crude two-layer filter whereby suspended solids are first subjected to a prefiltering effect of coarse media and then through high permeable coarse and fine to medium layers. This filter is advantageous due to the very low head loss development and simplicity in cleaning. It also has a higher storage capacity than a conventional sand filter.

The under-drainage consists of graded stone, as shown in Figure 7.3. The detailed design of the filter is given in Vigneswaran and Joshi (1990). The main filter comprises a 30-cm layer of sand and broken stone matrix filter (20–25 mm stones filled with fine sand of size 0.25–0.50 mm). The use of fine sand in the matrix filter helps to remove coliform organisms, which are prevalent in many rural groundwaters. The stone and sand are mixed in a 4:3 ratio and filled in layers to ensure uniform compaction. The filtration rate is controlled by a control valve in the clear water outlet. The optimum filtration velocity was 0.7 $m^3/m^2\cdot h$. Backwashing for 20–25 minutes at a rate of 3.5 $m^3/m^2\cdot h$ is sufficient to wash away any deposits in the filter.

From the pilot-scale runs with raw water containing 4.5–5.7 mg Fe/L and 1.3–1.6 mg Mn/L, the filtered water contained 0.6 mg Fe/L and 0.36 mg Mn/L, even after a run of 70 hours. The iron removal percentage varied from 85–97%, whereas the manganese removal varied from 80–100%. Higher removal with longer filter runs were possible with an increase in filter depth. The unit is also capable of removing turbidity of 50–70 NTU with an efficiency of 92–96%.

Removal of Manganese

When raw water contains manganese in addition to iron, the above-mentioned iron removal processes are practically ineffective for efficient removal of manganese. Precipitation in the form of hydroxide, or oxidation by oxygen, are generally feasible only at high pH (at least 9.0–9.5) values; oxidation by chlorine is sometimes possible, but only in the presence of large excess of chlorine, which then requires neutralization. The high residence time required for removal of manganese is also a problem.

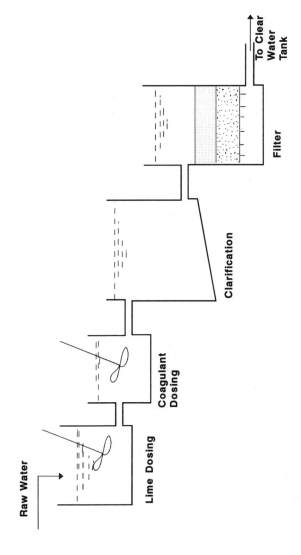

FIGURE 7.2 Iron removal plant for treatment of water with high iron content.

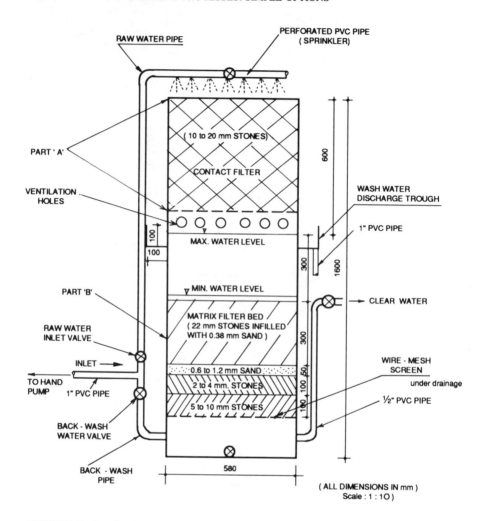

FIGURE 7.3   Details of the filter unit. (Vigneswaran and Joshi, 1990)

Sufficiently rapid oxidation can, however, be produced by chlorine dioxide, potassium permanganate, or ozone, which convert Mn^{++} to an oxidation level of +4 and precipitate it in the form of manganese dioxide. The reactions below show the mechanism of manganese removal:

$$Mn^{2+} + 2ClO_2 + 2H_2O \longrightarrow MnO_2 + 2O_2 + 2Cl^- + 4H^+ \qquad (7.2)$$

$$3Mn^{2+} + 2MnO_4^- + 2H_2O \longrightarrow 5MnO_2 + 4H^+ \qquad (7.3)$$

$$Mn^{2+} + O_3 + H_2O \longrightarrow MnO_2 + O_2 + 2H^+ \qquad (7.4)$$

According to these reactions, the theoretical quantities of oxidant required for each mg/L of Mn^{2+} are

Chlorine dioxide (ClO_2) = 2.5 mg/L
Potassium permanganate ($KMnO_4$) = 1.9 mg/L
Ozone (O_3) = 0.87 mg/L

The actual dose may differ from this considerably, depending primarily upon the pH, and also upon the contact time, existing deposits, organic matter content, etc. In practice, the dose generally ranges between 1–6 times the manganese content in the case of $KMnO_4$, 1.5–10 times for ClO_2, and 1.5–5 times for O_3. Figure 7.4 shows a schematic diagram of a gravity plant with concrete filter, in which iron removal can be combined with manganese removal, coagulation-filtration, and pH adjustment (Degremont, 1991).

Combined Manganese and Turbidity Removal

A pH of 9.5 is necessary for oxidation of manganese. Bratby (1988) noted that pH values close to 7 are required for turbidity removal. To accommodate these conflicting requirements, the normal sequence of chemical addition was inverted to chlorine–$FeCl_3$–lime. By adding $FeCl_3$ before pH adjustment to 9.5, the destabilization of suspended material occurred at the optimum pH. Also, the microflocs formed at the lower pH had a nucleation effect on subsequent manganese precipitation. The optimum time lag between $FeCl_3$ and lime addition is 15–30 seconds. Irrespective of the raw water quality, the optimum $FeCl_3$ dosage is 4 mg/L as $FeCl_3$.

Treatment Combined With Carbonate Removal

Chemical treatment with lime is an efficient and economical method for removal of free carbon dioxide, iron, and manganese. The dose should result in a final pH of 7.5 to 8.5 after sedimentation and filtration. The lime must be thoroughly mixed with the water. Aluminum sulfate (alum) may be used to assist flocculation and sedimentation. For removal of Mn, the pH should be above 9.

Carbonate removal with lime, which results in a high pH, produces favorable conditions for iron and manganese removal. At pH 8.2 almost all the ferrous carbonate precipitates, as does the ferrous hydroxide at pH 10.5. In presence of a high redox potential, Fe^{++} in solution may be precipitated in the form of $Fe(OH)_3$. The equation below shows the process:

$$Fe^{2+} + 3H_2O \longrightarrow Fe(OH)_3 + 3H^+ + e^- \qquad (7.5)$$

For manganese, the pH at which precipitation occurs is 9.2 in the case of carbonate and 11.5 in that of hydroxide. Partial carbonate removal, at a pH about 8,

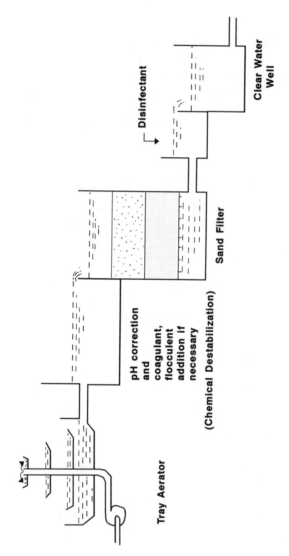

FIGURE 7.4 Iron and manganese removal in an open plant (using aeration and filtration).

may thus result in complete removal of iron. In some cases, such as catalytic carbonate removal plants, satisfactory manganese removal takes place at the same pH, whereas theoretically the process should be associated with total carbonate removal at pH 9.5 or 10. Figure 7.5 is a schematic diagram of a treatment plant employing partial carbonate removal, aeration, and filtration.

Potassium Permanganate-Manganese Greensand Filtration

Potassium permanganate is an effective oxidizing agent, particularly for manganese, which will precipitate when the pH is above 6.5. This application should be followed by filtration by sand, anthracite, or a catalytic filter containing iron and manganese oxides.

Potassium permanganate-manganese greensand filtration is similar to chlorination-filtration except for the differences in the oxidant and filter media. Zeolite or greensand must be first enriched with MnO_2, after which it acts as an electron exchanger. Manganese dioxide oxidizes Fe^{2+} and Mn^{2+}, which are then precipitated and retained by the filter media, while the MnO_2 is reduced to sesquioxide (Mn_2O_3). The manganese dioxide can then be regenerated, in a continuous or a batch process, by $KMnO_4$. However, the technique can only be used in small plants treating water with a low iron and manganese content and also free of organic matter. Pilot plant testing is recommended when using this method.

$KMnO_4$ is a very effective oxidant and has kinetic rates of oxidation that are relatively rapid. Among other benefits, it readily reacts with hydrogen sulfide, cyanides, phenols and other taste- and odor-producing compounds. However, excess $KMnO_4$ can increase soluble Mn levels and impart a pink color to the water if not properly controlled. Manganese greensand has the advantage that it can remove iron and manganese by a combination of sorption and oxidation (Lorenz et al., 1988).

The potassium permanganate-manganese greensand filtration technique has the following drawbacks:

1. high operating cost due to high chemical requirements
2. filter bed deterioration when pH falls below 7.1 (Wong, 1984)
3. the excess oxidant has to be removed

Because of these disadvantages, the chlorination-filtration technique is finding wider application at present.

Janda and Benesova (1988) proposed manganese removal in a fluidized bed reactor, using $KMnO_4$ as the oxidation agent. They noted that manganese was oxidized in the fluidized bed and the formed manganese oxides were separated and deposited as a compact layer on the media grains (specific gravity of the manganese oxide layer was 2.60). The manganese removal was 100% at the optimum $KMnO_4$ dose. The process was found to be efficient even when humates are present.

The oxidation of reduced Mn(II) by $KMnO_4$ and ClO_2 is extremely rapid. This method of removal has become popular during the last decade due to the emphasis

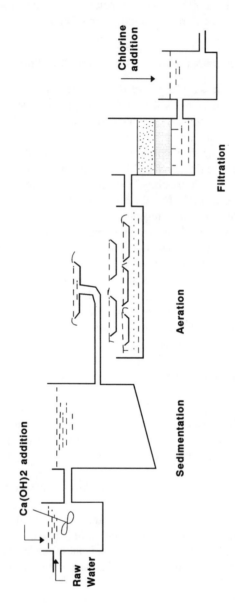

FIGURE 7.5 Diagram of partial carbonate removal, aeration, and filtration plant.

on organics removal and control of chlorination byproducts, which had limited the use of free chlorine as an oxidant. Although the application of alternative oxidants has increased in recent years, their higher costs compared to chlorine have forced many plants to significantly reduce the oxidant dosages. In organics-laden water, the competition between organic matter and the reduced metal for oxidant may result in poor Fe(II) and Mn(II) oxidation. It was, however, noted that the presence of significant amounts of humic and fulvic acids did not appear to cause complexation and the oxidation rate was not affected significantly (Knocke et al., 1991). A pH value of 5.5–6.5 is used in such waters (to ensure proper coagulation) to maximize organics removal. These pH values are not the optimal values for removal of manganese by oxidants, hence manganese removal under such conditions is not very efficient.

The experimental data for Fe(II) oxidation in the absence of organic matter indicated that the use of either $KMnO_4$ or ClO_2 resulted in extremely rapid oxidation rates. The complexed Fe(II) was very difficult to remove by the oxidation and subsequent precipitation of $Fe(OH)_3$. Efficient removal of complexed Fe(II) must rely on processes that can remove the organic matter, such as coagulation or adsorption onto powdered or granular activated carbon. Mn(II) oxidation rate increases as solution pH values increase. Low temperatures, (2–3°C) may cause problems in certain Mn(II) oxidation treatment schemes (Knocke et al., 1991).

The oxidant feed points should be properly placed in the treatment process. The Fe(II) oxidation is always faster than Mn(II) oxidation. Therefore chlorine should first be added to oxidize the Fe(II). After allowing sufficient contact time (e.g., 1 minute at pH 7), the $KMnO_4$ could be added for Mn(II) oxidation.

Chlorination-Filtration Technique

The system of chlorination-filtration is quite simple and consists of only a chemical feed system followed by filters. If pH adjustment becomes necessary, an additional small retention tank together with a pH adjustment system is incorporated into the system. Soda ash, caustic soda [NaOH] or lime [$Ca(OH)_2$] is used to adjust pH. Chlorine is used either in gaseous form or as hypochlorite. Table 7.2 depicts the chlorine requirements.

Table 7.2 Chlorine Requirements (Wong, 1984)

Impurity	Concentration of impurity (mg/L)	Chlorine required (mg/L)
Iron	1.0	0.62
Manganese	1.0	1.30

Since the oxidation is often slow in the case of manganese, the process is best suited for water with high iron concentration, unless a longer detention time can be provided. The filter can be a sand filter commonly used in the aeration-filtration process.

Parameters Influencing Removal Efficiency

The following parameters seem to influence the performance of the filter in the removal of iron:

1. amount of oxidant
2. pH
3. filtration rate
4. filter medium size
5. filter depth

In order to have a quick overview of the influence of the above parameters on iron removal, results of a laboratory-scale experimental study by Zainuddin (1985) are summarized in Table 7.3.

Table 7.3   The Effect of Related Parameters on Iron Removal (Vigneswaran et al., 1987)

Parameter	Effect
OCl⁻ dosage (used as NaOCl)	No significant improvement in removal when dosage was increased beyond 0.83 mg/L per mg/L of Fe^{++}. The change of dose did not give rise to significant change in head loss. Effluent quality was better at pH 9.5 than at pH 7.8 but head loss was higher at pH 9.5. The reason may be that gelatinous floc formed at higher pH caused blockage of filter pores. But a higher pH is necessary for effective manganese removal.
Filtration rate	At filtration rate of 5 $m^3/m^2 \cdot h$, the filter produced an effluent within standard (0.3 mg/L) for over 10 hours of run, whereas at high rates, such as 10 $m^3/m^2 \cdot h$, the effluent quality deteriorated beyond 0.3 mg/L after 5.5 hours run. Hence an optimum filtration rate must be chosen for each source of water supply based on lab-scale experimental studies.
Filter medium size	The larger the medium the more inferior the effluent quality but better in terms of head loss development. An optimum value for filter medium size has to be found from laboratory-scale studies.
Depth of filter	At filtration rate of 5 $m^3/m^2 \cdot h$, the effluent quality did not improve beyond a certain depth (e.g., 30 cm). But at higher filtration rates the effluent quality depends on depth, and higher filter depths were necessary to achieve an effluent quality that satisfied the required standard.

Application Status of Chlorination Filtration

The chlorination-filtration technique has not yet found wide application, but sufficient pilot-scale studies have been conducted to evaluate the effectiveness of this process. Russell (1977) has reported some of the results of his pilot study that prove the effectiveness of the chlorination-filtration process in removing iron and manganese. The process was found to be most effective at pH of more than or equal to 8.4. Filter media used were anthracite and sand. These media were found to be more economical than manganese greensand used in the permanganate-filtration method. These media also allowed the use of higher filtration rates without much pressure drop. This is due to their larger size (0.5–0.6 mm) than that of greensand (0.3 mm). The chemical costs were also found to be less. Russell (1977) also reported that the addition of extra chlorine was tolerable because it is a common disinfectant, but this is not the case with the addition of extra permanganate.

Two types of filter, dual and single media, were tested by the Marin Municipal Water District (MMWD) in Marin County, California. Table 7.4 shows the parameters used in the study. The results obtained were excellent, but with some differences between iron and manganese removal. Iron was satisfactorily removed with NaOH and Ca(OH)$_2$ as pH adjusting agents, but manganese was satisfactorily removed only when Ca(OH)$_2$ was used. Both chemicals gave satisfactory results after a few hours of run, may be due to the deposition in the filter pores that would have reduced the pore size. But this eventually caused higher pressure drops. Wong (1984) explained Ca(OH)$_2$ being effective on manganese removal in the following manner: At pH range of 8–8.5, MnO$_2$ colloids formed due to oxidation exhibit high negative surface charges. This MnO$_2$ was more effectively neutralized by divalent Ca^{++} than monovalent Na$^+$, so that the coagulation and flocculation of MnO$_2$ colloids were more effective. If the raw water is relatively hard (more than 75 mg/L as CaCO$_3$) then even NaOH could be used for pH adjustment. The raw water used in the above study had a hardness of 50 mg/L as CaCO$_3$ only.

Table 7.4 Parameters Used in Chlorination Filtration Pilot Study (Wong, 1984)

Parameters		Value
Filter media size	Dual media	300 mm anthracite (1.0–1.5 mm)
		300 mm silica sand (0.5–0.7 mm)
	Single medium	600 mm anthracite (1.0–1.5 mm)
Chemicals for pH adjustment		Caustic soda (NaOH), lime [Ca(OH)$_2$]
Iron and manganese in raw water		1–1.5 mg/L
Filtration rate		12 m^3/m$^2\cdot$h
pH		more than 8.5
Chlorine reaction time		20 minutes
Recommended limits for		
Iron		0.3 mg/L
Manganese		0.05 mg/L

The most common and readily available oxidant for manganese precipitation is chlorine. Other oxidants such as KMnO$_4$, although more efficient, are prohibitively expensive.

Chlorination-filtration seems to be a water treatment technique that could be readily and conveniently applied in developing countries. Most of the developing countries depend a great deal on groundwater for water supply, and iron and manganese are generally found in excess quantities in groundwater. In the treatment plants where aeration-filtration is used to remove iron and manganese, disinfection is always applied by postchlorination. If the prechlorination-filtration system is used, the postchlorination can be achieved with a smaller quantity of chlorine. From the discussion so far, it is clear that chlorination-filtration gives the combined effect of the above while eliminating the aeration stage. The advantages and disadvantages must be carefully studied at pilot scale before construction of industrial scale facilities. Both advantages and disadvantages of the chlorination-filtration process are discussed in Table 7.5.

Table 7.5   Advantages and Disadvantages of Chlorination-Filtration Process (Vigneswaran et al., 1987)

Advantages	Disadvantages
Cost effective	Chlorination may lead to the formation of chlorinated
Chlorination equipment is readily available	organics (THMs)
because of its wide usage in disinfection	pH and Ca++ must be carefully adjusted to prevent
Easy to operate and can be incorporated	scaling problems for consumers and in pipes
with disinfection	Large-scale use is still too limited to evaluate its
	effectiveness in large-scale application

Ion Exchange

Ion exchange should be considered only for the removal of small quantities of iron and manganese. There is a risk of rapid clogging, and ion exchangers also remove other cations, such as Ca, Mg, etc. They are not generally used to deal with specific iron and manganese problems, as they are not cheap and can only rarely be acceptable for industrial-scale treatment of drinking water.

Ozone and Activated Carbon

Ozone is sometimes used as an oxidant, in which case ozone consumption is kept to a minimum by preaerating the raw water (spray or cascade aeration). Aeration takes place in the ozonized atmosphere of the upper section of the ozone chamber (Mouchet, 1992). When the raw water contains any pollution, taste or odor, ozone-treated water can be filtered on a bed of activated carbon. These treatment methods are expensive and should be used only in the absence of other alternatives.

Effects of Lime Addition on Manganese Removal

The application of an oxidant and alum is effective in removing manganese by sand filtration. The alum dosage can be substantially reduced by the application of lime, with no effect on manganese removal. This phenomenon is also true for turbidity removal by alum-calcium treatment and by alum alone (Jenkins et al., 1984). The addition of lime was seen to be very effective, probably because of the dual role it plays, namely to increase the pH and provide Ca^{++}, both of which increase Mn removal efficiency.

In another study, Mn removal by filtration using different coarse media, at different pH, was compared in an attempt to develop a simple and cheap method of removal of Mn (Aziz and Smith, 1992). Among the filter media studied, were limestone, gravel, crushed brick, and without solid media. It was seen that the highest removal efficiency (95%) could be obtained with limestone at pH 8.5, indicating that the presence of $CaCO_3$ is beneficial. Comparison of limestone in solution and limestone as a rough media showed that the rough media had an advantage. Comparison of synthetic $CaCl_2$ and $CaCO_3$ solutions indicated that it was the CO_3 species that contributed to the Mn removal.

Drawbacks of Conventional Processes

Mouchet (1992) identified the following problems associated with iron and manganese removal using the above-mentioned processes. The poor functioning of treatment plants may be due to one or several of the following reasons.

- Iron complexation: 75% of the cases (the complexing agent is silica; the other 25% is due to humic substances)
- Oxidation pH being too low: 35% of cases
- Negative effect of chlorination (inhibition of biological phenomena that would otherwise enhance treatment): 25% of the cases
- Problems related to flocculation of oxidized iron or manganese: 20% of the cases
- Effective size of the filter sand being too large: 20% of the cases
- Lack of analytical data or representative samples at the time of design: 20% of the cases
- Deterioration of the raw water quality over time: 15% of the cases
- Interference by nitrification: 15% of the cases
- Oxidation time being too short: 10% of the cases
- Reagent dosing locations being not optimum: 10% of the cases

The main problems with these processes are now known and have been overcome. Yet these processes have limitations. Filtration rates are restricted to maximum of $10–15 \ m^3/m^2{\cdot}h$, a clarification stage is required when the iron content is $>5–10 \ mg/L$, chemical treatment is indispensable when complexing phenomena occur, and the retention capacity between two backwashes, at $0.2–1.2 \ kg \ Fe/m^2$ and $0.1–0.7 \ kg \ Mn/m^2$, is low.

Biological Treatment

The metabolism of certain autotrophic microorganisms is based on the oxidation of iron and manganese. If conditions are favorable, reactions are very rapid and the two elements can completely be removed. As ammonia nitrogen is essential to bacterial feeding, the presence of ammonia in raw water provides the favorable conditions for this type of treatment. The principle does not differ essentially from that of the oxidation-filtration processes, but the design of the units has certain special features relating to oxygen content, filtration rate, and effective size of sand.

The biological technique is widely used in Holland and Germany, and is advantageous primarily when water simultaneously contains iron, manganese, and ammonia. In Holland, excellent results were obtained using dry filters (a type of trickling filter used in drinking water treatment) to perform simultaneous oxidation and filtration. The boundary line between purely chemical oxidation and biological treatment cannot be clearly defined, and many plants utilizing physicochemical treatment processes for Fe and Mn removal in fact owe a fraction of their efficiency to the activity of microorganisms (Degremont, 1991).

Iron was the first element for which biological removal techniques were discovered and implemented. The bacteria involved in biological iron removal have the unique property of causing oxidation and precipitation of dissolved iron under pH and redox potential conditions that are intermediate between those of natural ground-water and those required for conventional (physicochemical) iron removal. These microorganisms accelerate Fe(II) oxidation through their catalytic action and then accumulate the products of this oxidation in the form of a substance far less likely to clog filters. The precipitates that accumulate around bacterial cells, sheaths, stalks, and polymer filaments, form sludge that is denser, less likely to clog filters, and easier to thicken and dewater where applicable (Mouchet, 1992).

The biological treatment is heavily dependent on the oxygen admission levels. Severe restrictions on oxygen levels (generally from 5 to 25 µg/L) limit the activity of iron bacteria in filters. The biological treatment provides the answer to the frequent problem of complexing. When the iron is chelated by organics, any of the several types of heterotrophic iron bacteria is capable of utilizing the organic fraction of the complex, thereby releasing the iron that can be oxidized by catalysis. However, the contact time allowed by the treatment process should be sufficient for these reactions to be completed.

In some of the existing French plants, physicochemical treatment has been converted to biological treatment through one or several of measures such as modifying the pH, changing aeration conditions, eliminating a strong oxidant or, more generally, any other reagents added at the head of the treatment line, postponing chlorination until the end of the treatment line, or changing the filter media. These simple modifications, when performed under appropriate conditions, improved treatment efficiency in terms of product water quality, operation costs (longer filter runs, decreased labor requirement, savings on backwash water, partial or total reduction in the use of costly reagents), or both. In France about 20 conventional treatment plants have now been adapted to biological processes (see Table 7.6).

Table 7.6   Examples of Biological Fe and Mn Removal Plants in France (Mouchet, 1992)

Name	Capacity (m³/h)	Filtration rate of first stage (biological iron removal)(m³/m²·h)	Filtration rate of second stage (manganese removal)(m³/m²·h)	Type of manganese removal
Hochfelden	600	37	10	Physicochemical, then converted to biological
Ramelshausen	150	40	21	Physicochemical, then converted to biological
Blaye	120	39	24	Biological
Bischwiller	60	24	16	Physicochemical

Adapting plant design to actual raw water quality dispenses with the need for pH adjustment (except to correct the carbonate balance of the treated water, if necessary). Also, biological treatment eliminates the problem of iron sequestration by silica and flocculation of the iron precipitate. However, the water should be chlorinated before discharge into the network. Unlike conventional processes in which clarification is considered necessary (when the raw water iron content exceeds 5–10 mg/L), biological iron removal processes allow for direct filtration, even when the raw water iron content is very high (in practice up to 25 to 30 mg/L).

Taking into account the installations common to both types of facility, the capital cost of a biological iron removal plant amounts to approximately 60% of that of a conventional plant. Operating costs are also considerably lower for a biological plant.

Biological iron removal is ideal for water with an acidic or neutral pH as well as high iron and silica contents, and that is devoid of toxic substances such as H_2S, heavy metals, and hydrocarbons. The water applied to the filters should not contain more than 0.01 mg H_2S/L. Iron removal rates of only 50% have been reported with 0.45 mg Zn/L, with total inhibition of treatment when zinc concentration reaches 1 mg Zn/L. When ammonia is present in the raw water, a high iron content may be effectively removed by biological iron removal just upstream of the actual nitrification stage.

Manganese can be biologically oxidized to be deposited as MnO_2, in the form of a black precipitate coating the free cells or as pustules on the sheaths of the filamentous bacteria. To oxidize manganese, the bacteria need more stringent conditions than for iron oxidation; in particular, they need an aerobic environment. Generally, it is not possible to achieve simultaneous removal of Fe^{2+} and Mn^{2+} in a single reactor, except with extremely low filtration rates.

Biological processes now offer the best alternative to conventional plants due to the following advantages (Mouchet, 1992):

- these processes can be operated at high filtration rates (up to 50 $m^3/m^2 \cdot h$)
- coarse sand media 0.95–1.35 mm or larger can be utilized, which give low head loss and higher retention capacity (1–5 kg Fe or Mn/m^2)
- elimination of chemical reagents
- flexibility of operation
- reduced capital and operating costs
- good sludge treatability

The limitations of biological processes are the need for two filtration stages in the event of simultaneous presence of iron and manganese, interference by certain inhibiting substances (H_2S, Zn), and the necessity of NH_4^+-N.

Biological iron and manganese removal plants already exist in France (more than 100), Germany, Belgium, Bulgaria, Finland (Seppanen, 1992), Netherlands, Togo, and Argentina. But implementation of this biological process demands a high degree of expertise, as well as preliminary testing.

7.1.4 Conclusion

Iron and manganese generally occur in groundwater in the dissolved, reduced form. Water treatment for their removal usually includes an aeration step to form insoluble precipitates. Oxidation is achieved by aeration or chemical oxidants. Aeration has been shown to be very successful for removal of iron above pH 6.5, while for Mn, it is effective only above pH 9.5. In addition, a contact time of up to one hour after aeration may be required to form Mn floc.

Chlorine, ozone, chlorine dioxide, and potassium permanganate are the main chemical oxidants used. Chlorine is effective and cheapest; however, the potential formation of THMs has reduced its use. The other three are extremely effective, but their relatively high cost has curtailed their use. Lime or lime/alum addition has been found to be an extremely effective method, with relatively low cost. Complexation or sequestration is another form of treatment that is low cost, but it does not remove the constituents, only keeps them in solution, and is therefore practiced only in controlled uses (Robinson et al., 1992).

7.2 DEFLUORIDATION

7.2.1 Introduction

Excessive amounts of fluoride in water may affect dental hygiene, causing dental and skeletal fluorosis; hence, it should be reduced to an optimum level. Health authorities prescribe the range for desirable and allowable fluoride concentration in raw water as 1.0 and 2.0 mg/L, respectively. Excessive fluoride concentration was found to cause badly mottled teeth in children (Hyde, 1985). Very high levels of fluoride can also cause dental malformation, stained enamel, decalcification, mineralization of tendons, digestive and nervous disorders, and skeletal fluorosis. On the other hand, a small quantity of fluorine in drinking water (0.4–1.0 mg/L) is found to be beneficial, promoting the formation of dental enamel and protecting teeth against caries (tooth decay). Therefore, it is quite important to treat fluoride to achieve the optimum level. The upper limits recommended by the US EPA vary from 0.9–1.7 mg/L, an average value depending upon the climatic conditions of the usage area. The permissible content falls as the mean annual temperature rises (Table 7.7). The lower values pertain to high-temperature areas where water consumption by children is higher. The EEC recommended value for the fluorides is 0.7–1.7 mg/L.

Table 7.7 United States Public Health Service Recommended Limits of Fluoride Concentration

Annual average of maximum daily air temperature (°C)	Recommended control limits (mg F/L)		
	Lower	Optimum	Upper
10.0–12.1	0.9	1.2	1.7
12.2–14.6	0.8	1.1	1.5
14.7–17.7	0.8	1.0	1.3
17.8–21.4	0.7	0.9	1.2
21.5–26.2	0.7	0.8	1.0
26.3–32.5	0.6	0.7	0.8

Fluorides are mostly found in groundwater sources, and the concentration ranges from one to hundreds of mg/L. Water with high fluoride content is usually found at the foot of high mountains and in areas with certain geological formations. In seawater, fluoride is found up to 1.35 mg/L, while in surface water the concentration is low (less than 0.3 mg/L) and generally does not pose any health problem. However, in areas with fluoride-rich volcanic rocks, extremely high fluoride concentrations (greater than 1000 mg/L) have been found in surface water (Choi and Chen 1979). Since fluoride, as it occurs in drinking water, is colorless, odorless, and tasteless, its presence in excess only becomes evident if chemical analysis is performed or if cases of dental or skeletal fluorosis are found.

The fluoride in drinking water may be either due to natural occurrence or to industrial activities. The latter includes various industrial processes that lead to addition of fluorine in surface and subsurface water sources. These industrial processes are electroplating, manufacture of aluminum, glass, electronic components, pesticides, disinfectants, wood preservatives, metals, etc. Fertilizer production utilizing phosphate rocks, and extensive use of phosphate fertilizer also introduces large amounts of fluoride in water.

7.2.2 Treatment Techniques

For other water treatment methods, the objective is to remove the impurity to the maximum level possible, but defluoridation is special in that the treated water should have an optimum average fluoride level as shown in Table 7.7. The standard methods of fluoride removal are (1) coagulation and co-precipitation followed by sedimentation and (2) adsorption.

Alum (aluminum sulfate), lime, and magnesium compounds such as dolomite are the commonly used coagulants. Calcium phosphate is used as the co-precipitant. Activated carbon, bone char, ion exchange resins, synthetic zeolite, and activated alumina are the common adsorbents for fluoride removal (Hao and Huang, 1986). In the adsorption methods, various granular insoluble media remove the fluoride as the water percolates through them. Fluoride is removed by ion exchange or by chemical reaction with the adsorbent. These media can be periodically regenerated by chemical treatment.

Fluorides can be chemically precipitated by using multivalent metal ions. Fluoride ion content in a water supply can be reduced from 3.6 mg/L to 0.25 mg/L by using 225 mg/L of alum at pH 6.5–7.5. Whichever the method used, the raw water alkalinity has a significant effect on the chemical dosage used for defluoridation.

Precipitation Methods of Fluoride Removal

Fluorine has a high affinity to calcium and calcium salts such as calcium hydroxide [$(CaOH)_2$], calcium sulfate [$CaSO_4$] and calcium chloride [$CaCl_2$], which are used for precipitating fluoride present in water by converting it to insoluble calcium fluoride [CaF_2].

$$Ca(OH)_2 + 2HF \longrightarrow CaF_2 + 2H_2O \qquad (7.6)$$

The drawbacks of the precipitation method include the necessity of several reagents to reduce fluorine to less than 1 mg/L, higher shipment and treatment costs, and the large volume of sludge produced.

Lime Softening

Lime is the cheapest chemical employed for the removal of fluoride. But it is impossible to reduce the fluoride level to 1 mg/L using only lime, as the solubility of the precipitate formed (CaF_2) is about 7.7 mg/L as F (Choi and Chen, 1979). Rabosky and Miller (1974) showed that fluoride removal by chemical method of lime precipitation was very difficult when fluoride concentration was below 20 mg/L. In other words, removal by lime is ineffective for dilute solutions of fluoride.

Lime softening can be used when the treated water has sufficient magnesium content, since the magnesia absorbs and precipitates the fluoride. Otherwise the water has to be enriched with magnesium by addition of magnesium sulfate or using dolomitic lime. About 50 mg/L of magnesium is required to remove 1 mg/L of fluorine. Bulusu (1984) reported the use of magnesia for defluoridation in detail and showed that large doses are necessary. Also, since the pH of treated water exceeded 10, pH adjustment by acidification or recarbonation was unavoidable.

Removal by Alum Coagulation

The principle behind this method is to produce aluminum hydroxide in water, which acts as an adsorbing surface for fluorine. Alum reacts with the alkalinity [$Ca(HCO_3)_2$] in water to produce aluminum hydroxide [$Al(OH)_3$]. The fluoride gets adsorbed on the surface of aluminum hydroxide and is removed by sedimentation along with aluminum hydroxide precipitate.

This method requires a very large dosage of alum, ranging from 150 to 1000 mg/L according to the circumstances. The treated water may contain a large amount of dissolved aluminum, which has to be reflocculated after pH adjustment. Since the alum doses required for fluoride removal are much higher than those commonly used for turbidity and color removal, it was not considered feasible until the Nalgonda technique (described in Section 7.2.3) was introduced.

Aluminum Chloride

Aluminum chloride can be applied to water as a defluoridating chemical either alone or in combination with aluminum sulfate (Bulusu, 1984). This is useful especially when the use of alum is constrained by the limits of sulfate in potable water. (Allowable limit is 400 mg/L for sulfate and 600 mg/L for chloride. These limits correspond to 1000 mg/L aluminum sulfate or 750 mg/L aluminum chloride.) Due to this, it is not desirable to exceed doses of 1000 mg/L alum and 700 mg/L aluminum chloride (anhydrous). When a combination of these chemicals has to be used due to unfavorable raw water alkalinity and fluorides, it is preferable not to exceed a dose combination of 700 mg/L aluminum sulfate and 700 mg/L aluminum chloride to prevent adverse effects on potability of treated water. Experimental investigations (Bulusu, 1984) reveal that it is possible to defluoridate all naturally occurring fluoride waters by suitably using aluminum sulfate or aluminum chloride or a combination of these two. The doses of aluminum salts are dependent on the alkalinity of water, with high alkalinities needing higher dosages.

Ion Exchange and Adsorption Methods of Fluoride Removal

The adsorption method is economically more favorable for low concentrations of fluoride in water or for relatively low concentrations of fluoride after the removal of fluorides by precipitation methods to the 10–20 mg/L level (Choi and Chen, 1979). Synthetic ion exchange resin media, activated alumina, tricalcium phosphate, or activated carbon can be used for defluoridation by ion exchange.

Defluoridation Using Activated Alumina

Activated alumina as an adsorbent of fluoride is the most popular and most effective defluoridating method due to its ease of application and cost-effectiveness. Water is allowed to percolate through a bed of granular media of grain size 0.3–0.5 mm (i.e., filtration using activated alumina as medium). Rate of adsorption from the solution is strongly dependent on the particle size of the adsorbent. As the basicity of the raw water increases, the effectiveness of removal is reduced. The removal also decreases with the number of filter cycles. Bed regeneration is necessary with alum or with weak caustic soda followed by neutralization using sulfuric acid.

The basic principles of fluoride removal technology utilizing activated alumina are (Rubel and Wooseley, 1979):

1. Optimizing the environment for sorbing the fluoride ions to activated alumina surfaces.
2. Preventing the competing ions from occupying alumina surfaces that should be reserved for fluoride ions.
3. Upon regeneration of an expanded bed, taking all steps necessary to remove all fluoride ions from the bed before returning it to treatment.

There are four steps in the operation cycle, namely, treatment, backwash, regeneration, and neutralization. These modes are discussed in detail by Rubel and Wooseley (1979).

The fluoride removal by activated alumina is influenced by pH, surface loading (the ratio of total fluoride concentration to activated alumina dosage), and presence of common interfering anions such as alkalinity (HCO_3^-) and sulfate. The optimum pH value changes with the type of adsorbent used. A value of pH = 5 was noted for fluoride removal by activated alumina. The smaller activated alumina particles (24×48 mesh) exhibit higher removal efficiency than the larger alumina particles

(8 × 14) mesh. This phenomenon is due to the greater specific surface area of the smaller adsorbent size (Hao and Huang, 1986).

Activated Bauxite

The main component of activated bauxite is also Al_2O_3. It has other oxides such as Fe_2O_3, TiO_2, and SiO_2 complexes (red mud) in appreciable amounts. Because of this, the physicochemical characteristics of activated bauxite that affect adsorption will differ from those of activated alumina (Choi and Chen, 1979). The removal efficiencies of both activated bauxite and activated alumina decline sharply outside the optimum (near neutral) pH range. This indicates that both can be regenerated easily with either acid or alkali.

Activated carbon is distinctly inferior to both activated alumina and activated bauxite for fluoride removal. Activated alumina is superior to activated bauxite due to a wider optimum pH range, higher removal capacity, lower water and mechanical wear, and lack of leachable impurities. Activated bauxite is much cheaper than activated alumina; therefore, even if the activated bauxite is inferior to activated alumina in performance, it is competitive due to the cost factor. The optimum pH range for maximum or near-maximum removal of fluoride by activated bauxite and activated alumina is 5.5–7.0 and 5.0–8.0, respectively.

The efficiency of fluoride removal generally increases as the initial concentration of fluoride in solution is decreased. To achieve a final fluoride concentration of less than 1 mg/L requires an initial fluoride concentration lower than 40 mg/L, consuming an adsorbent dosage of 25 g/L. Fluoride concentrations above 40 mg/L are rarely found, hence adsorption using activated alumina or activated bauxite is quite adequate.

The presence of other chemical species does not seriously affect the fluoride removal when activated alumina or activated bauxite is used. Nevertheless, the pH of the suspension is the most critical factor in determining the fluoride removal efficiency.

Fly Ash

Investigations of the use of fly ash as an adsorbent in defluoridation of water and wastewater have indicated that it has high potential as a simple and economical methodology (Chaturvedi et al., 1990).

Other Processes of Fluoride Removal

The fluoride and other excess minerals can be removed from the water using reverse osmosis. Electrochemical defluoridation processes that use aluminum anodes are also available. Some of the limitations of these methods are

- high capital cost
- high operation and maintenance costs

- low capacity for removing fluoride
- lack of selectivity for fluorides
- undesirable effects on water quality
- separation or replacement problems

7.2.3 Alternative Defluoridation Techniques for Developing Countries

Nalgonda Technique

The Nalgonda technique uses the aluminum sulfate treatment to effectively remove fluoride. It involves the addition of lime or sodium aluminate and aluminum sulfate in sequence followed by flocculation, sedimentation, filtration, and disinfection. Extensive research produced the Nalgonda technique in India, where fluorine was a big problem in several water sources (Bulusu and Pathak, 1974; Nawlakhe et al., 1974). This technique has been satisfactorily tested and proved in the laboratory as well as in the field. However, when evaluated by the University of North Carolina (UNC), it was found to be relatively inefficient in terms of chemical consumption, and an improved method was developed in laboratory studies at UNC (Robinson, 1984). The process is quite simple and requires quite unsophisticated equipment and materials. A typical fill and draw type defluoridation plant is shown in Figure 7.6.

Sodium aluminate or lime is added first into the water while stirring. The addition of bleaching powder and alum follows. The flocs formed are allowed to settle. The subsequent filtration produces the consumable water. Filtering may not be necessary in certain cases where flocs settle well. Bleaching powder provides disinfection. Lime was found to be cheaper than sodium aluminate. The quantity of lime necessary was found to be only around 4–5% of that of alum.

Alkalinity is a significant factor in the removal of fluoride. Results of an experimental study (Nawlakhe et al., 1975) on the effect of alkalinity in fluoride removal is shown in Figure 7.7. Here the parameter Z denotes the weight of fluoride removed by a unit weight of alum. Figure 7.7 shows that the removal is dependent upon the initial fluoride levels as well. At a low initial fluoride level, low alkalinities are effective, whereas at high initial levels relatively higher alkalinities are required.

Increased alum dose also produces better fluoride removal. But there is no definite relationship between alum dose and fluoride removal, and experimental studies may be required for the water to be treated. Since the alkalinity has a significant effect on removal, the addition of lime must be carefully controlled.

Design Concepts

There are no mathematical formulations in the Nalgonda design. In India, the fill and draw type, as shown in Figure 7.6, is popular (NEERI, 1978), and its design is based on past experience. The plant is suitable for a community with a population of 200 to 2000. The plant consists of a hopper bottom cylindrical tank. The depth of the tank

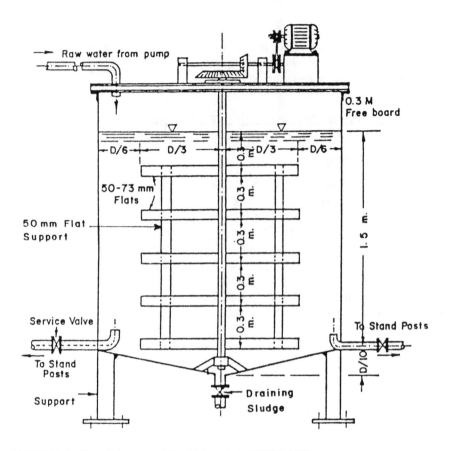

FIGURE 7.6 Fill and draw type defluoridation plant. (NEERI, 1978)

is generally 2 m and the diameter varies with the capacity. Typical design dimensions of the plant are given in Table 7.8.

A motor rotates the paddle used for mixing and flocculation with a speed of 60 to 80 rpm. For smaller plants, the paddles may be rotated manually. The stirring time is 10 minutes and settling follows for 1–2 hours. Sludge settles to the bottom and the supernatant is decanted as shown. Sludge is withdrawn from the bottom. This particular type does not require filtration, but filtration may be provided depending on the effluent clarity. The plant is operated on the batch mode with one operation taking 4 hours. Studies prove that the continuous mode of operation is also feasible (Nawlakhe et al., 1975). Figure 7.8 gives a flow diagram of a continuous-mode operation.

Relative Merits

The advantages and disadvantages of this process are summarized in Table 7.9.

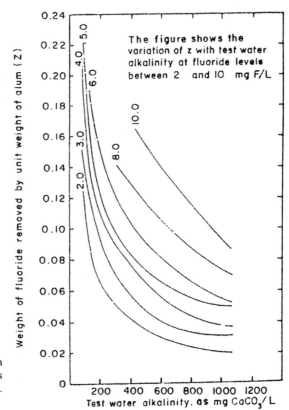

The figure shows the variation of z with test water alkalinity at fluoride levels between 2 and 10 mg F/L

FIGURE 7.7   Relationship between test water alkalinity and the fluorides removed by unit quantity of alum. (Nawlakhe et al., 1975)

Table 7.8   Plant Dimensions for Population Range from 200 to 2000 (NEERI, 1974)

Population	Volume (m³)	Diameter D (m)	Water depth (m)	Free board (m)	Depth of sludge cone (m)	Shaft diameter (mm)	H.P. required
2000	30.0	5.0	1.5	0.3	0.50	50	5.0
1500	22.5	4.5	1.5	0.3	0.45	50	5.0
1000	15.0	3.6	1.5	0.3	0.36	50	4.0
500	7.5	2.5	1.5	0.3	0.25	50	3.0
200	3.0	1.7	1.5	0.3	0.17	50	3.0

Table 7.9   Merits and Drawbacks of Fill and Draw Type Defluoridation

Advantages	Disadvantages
Simple technique and needs little maintenance	There are more parameters requiring careful control for proper removal
Semi-skilled labor is sufficient	
Does not need complicated equipment and thus no spare parts problems	The settling takes more time than the conventional turbidity removal
Operation is sufficiently flexible to suit the variable situation	The treatment mechanism is not very well understood
Incorporates disinfection as well	Requires substantially higher alum dosage than that for turbidity removal
Can be constructed totally out of locally available materials	

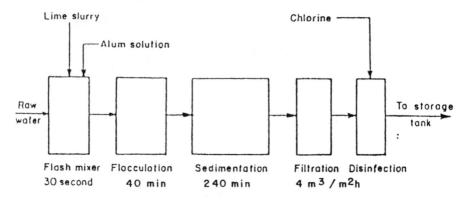

FIGURE 7.8   Flowsheet for defluoridation of water by Nalgonda technique. (Nawlakhe et al., 1975)

Defluoridator for Individual Households

A cheap and simple deflouridator has been developed in Thailand (Phantumvanit et al., 1988) that reduces the fluoride content and provides water that is clean, colorless, and odorless and having an improved taste. The main objectives of this device is

- low capital cost and low operation and maintenance cost
- simple design
- active ingredients that can be changed by the users themselves
- capacity to reduce fluoride content from 5 mg/L to 0.5 mg/L
- improvement of general quality of water
- to have ingredients that maintain their activity for an acceptable period of time

This defluoridator is based on the filtration and adsorption principle and uses charcoal and charred bone meal. As shown in Figure 7.9, it consists of a container and a filter. The container is made of a piece of PVC pipe, 75 cm long and 9 cm in diameter, with an outlet tap at the bottom and a cap with a small hole for water intake at the top. The filter comprises a bottom layer of 300 g crushed charcoal, mainly for absorption of color and odor, a middle layer of 1000 g charred bone meal, which has high fluoride adsorption capacity, and a top layer of approximately 200 g clean pebbles to prevent the intermediate layer from floating. The middle layer is prepared using bone meal of 40–60 mesh size and is activated by heating in an electric furnace to 600°C for 20 minutes.

In order to assemble the filter column, 3000 g of crushed charcoal is placed in the bottom of a plastic bag measuring 10–12 cm in diameter and 80 cm in length. The bag should be slightly wider than the PVC container, to enable it to be folded over the rim ensuring that raw water does not come into contact with treated water. A hole is cut at each of the two corners at the base of the bag for the treated water to pass out. The filter bag is placed in the container and 1000 g of charred bone meal followed by 200 g of pebbles are poured into the bag. A hole is then made at the level of the tap. Well water, usually stored in a clay jar, is siphoned to the top of the

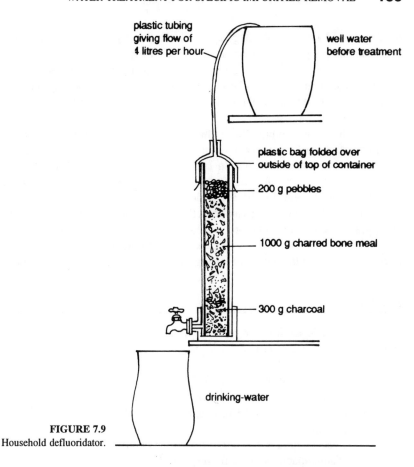

plastic tubing
giving flow of
4 litres per hour

well water
before treatment

plastic bag folded over
outside of top of container

200 g pebbles

1000 g charred bone meal

300 g charcoal

drinking-water

FIGURE 7.9
Household defluoridator.

defluoridator by means of a small plastic tube and a flow rate of 4 L/h is obtained. The defluoridated water is collected in another jar directly under the tap.

In laboratory-scale studies, with a flow rate of 4 L/h, the defluoridator reduced the fluoride content of 480 L of water from 5 to less than 1 mg/L. The first 20 liters of filtered water is normally discarded, after which the water is clean, odorless, and ready to be used for drinking or cooking. The filter remains active for 1–3 months, depending on the initial fluoride level and the amount of water consumed. The daily handling of the defluoridator and the periodic changing of the filter can be done by the users themselves.

7.2.4 Conclusion

The drawbacks of defluoridation by precipitation are the requirements of additional reagents, higher treatment costs, and a large volume of sludge generation. Alum is not economically feasible in some cases due to high dosage requirements. A more economical form of aluminum is calcined aluminum oxide (activated alumina), which has a fluoride adsorption value of approximately 1.43 mg F/g of alumina with a mesh size of 28 to 48 and a density of 800 g/L. It can be readily regenerated and reused. Adsorption methods are generally appropriate for relatively low concentrations of fluoride after the removal of fluorides by precipitation methods to the 10–20 mg/L level. Nevertheless, attention should be paid to avoid possible groundwater contamination resulting from the discharge of high fluoride-containing wastewater from the backwash and regeneration processes.

The Nalgonda technique for defluoridation is well suited for small to medium-sized communities commonly found in developing countries. Its simplicity and low cost make it a viable technique in India and NEERI of India has done extensive research to testify its value. There are simple design guidelines and the plant can be constructed easily with locally available materials.

Domestic drinking water can be defluoridized using low-cost multimedia filters using gravel, bone char, and charcoal as the filter media. This method is good only for single-household drinking water supplies, as the production rate is limited by the high retention times and the necessity for regeneration or replacement of the bone char.

EXAMPLE

Design a cascade aerator for removal of iron from raw groundwater, given the following design guidelines.

- Number of drops = 4–6
- Height of a drop = 30–60 cm
- Overflow rate = 0.01 m³/s over 1 m length of step
- Height of aerator = 2–3 m
- Cascade area = 1.5–2.0 m²/(m³/min) of flow
- Plant capacity = 5000 m³/day

Solution

Plant flow rate = 5000/(24 × 60) = 3.47 m³/min. Taking cascade area = 2.0 m²/(m³/min) of flow,

$$\text{required area} = 3.47 \times 2 \text{ m}^2$$
$$= 6.94 \text{ m}^2$$

Assume a circular aerator, as given in Figure 7.10. d_3 is the diameter of the bottom step.

$$(\pi/4) \, d_3^2 = 6.94 \tag{7.7}$$

Therefore, $d_3 = 2.97 \approx 3$ m (say). Limiting the step width to 0.4 m to ensure that the water jumps to the next step,

$d_2 = 2.2$ m

$$(7.8)$$

$d_1 = 1.4$ m

Total height of the drop = $0.5 \times 4 = 2.0$ m. Perimeter of the lower step = $\pi d_3 = \pi \times 3$ = 9.42 m, therefore,

Overflow rate = $\dfrac{3.47 \text{ m}^3/\text{min}}{9.42 \text{ m}}$
  = 0.37 m³/m·min

Allowable overflow rate = 0.6 m³/m·min

$$0.37 < 0.6 \qquad\qquad (7.9)$$

Therefore, the selected values are suitable.

Design of the Inlet Pipe

Assuming a flow velocity of 1.2 m/s = 72 m/min.

Cross-sectional area of inlet pipe = Q/v
  = (3.47 m³/min)/(72 m/min)
  = 0.0482 m²

$(\pi/4)$ d² = 0.0482

d = 247 mm ≈ 250 mm (say)

The rise of water jet upon leaving the inlet pipe,

h = v²/2 g
  = (1.2)²/2 × 9.81
  = 0.073 m
  = 73 mm

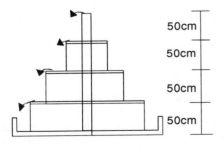
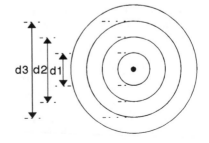

FIGURE 7.10 Cascade aerator.

REFERENCES

AWWA Trace Organics Substances Committee, Committee report on research needs for the treatment of iron and manganese, *J. AWWA*, 79, 119, 1987.

Aziz, H. A. and Smith, P. G., The influence of pH and coarse media on manganese precipitation from water, *Water Research*, 26, 853, 1992.

Bratby, J. R., Optimizing manganese removal and washwater recovery at a direct filtration plant in Brazil, *J. AWWA*, 80, 71, 1988.

Bulusu, K. R., Defluoridating of waters using combination of aluminum chloride and aluminum sulphate, *J. Institute of Engineers of India, Environmental Engineering*, 65(Oct.), 22, 1984.

Bulusu, K. R. and Pathak, B. N., Fluoride, its effect on human health and defluoridation of water, methods and their limitations, Paper Presented at the Symposium of Fluorosis, G.S.I., Hyderabad, India, 1974.

Chaturvedi, A. K., Yadava, K. P., Pathak, K. C. and Singh, V. N., Defluoridation of water by adsorption on fly ash, *Water, Air and Soil Pollution*, 49, 51, 1990.

Choi, W. W. and Chen, K. Y., The removal of fluoride from waters by adsorption, *J. AWWA*, 71, 562, 1979.

Degremont Handbook on Water Treatment, Vols. I and II, Lavoisier Publishers, Paris, 1991.

Hao, O. J. and Huang, C. P., Adsorption characteristics of fluoride on to hydrous alumina, *J. Environmental Engineering Division, ASCE*, 112, 1054, 1986.

Hyde, M., *Treatment of Groundwater, Developing Worldwater*, Grosvenor Press International, New York, 1985.

Janda, V. and Benesova, L., Removal of manganese from water in fluidized bed, *AQUA*, 37, 313, 1988.

Jenkins, J. R., Benefield, L., Keal, M. J. and Peacock, R. S., Effective manganese removal using lime as an additive, *J. AWWA*, 76, 82, 1984.

Knocke, W. R., Benschoten, J. E. V., Kearney, M. J., Soborski, A. W. and Reckhow, D. A., Kinetics of manganese and iron oxidation by potassium permanganate and chlorine dioxide, *J. AWWA*, 83, 80, 1991.

Lorenz, W., Seifert, K. and Kasch, O. K., Iron and manganese removal from ground water supply, *Public Works*, 11(9), 57, 1988.

Mouchet, P., From conventional to biological removal of iron and manganese in France, *J. AWWA*, 84, 158, 1992.

Nawlakhe, W. G., Kulkarni, D. N., Pathak, B. N. and Bulusu, K. R., Defluoridation of water with alum, *Indian J. Environmental Health*, 16, 1, 1974.

Nawlakhe, W. G., Kulkarni, D. N., Pathak, B. N. and Bulusu, K. R., Defluorination of water by Nalgonda Technique, *Indian J. Environmental Health*, 17, 26, 1975.

NEERI, Nalgonda technique for defluoridation, fill and draw type defluoridation plant, *Technical Digest*, No. 59, January 1978.

NEERI, Defluoridation of water by Nalgonda technique, *Technical Digest*, No. 46, October 1974.

Phantumvanit, P., Songpaisan, Y. and Moller, I. J., A defluoridator for individual households, *World Health Forum*, 9(4), 555, 1988.

Rabosky, J. G. and Miller, J. P., Fluoride removal by lime precipitation and alum and polyelectrolyte coagulation, *Proceedings of the 29th Purdue Industrial Waste Conference*, 670, 1974.

Robinson, G., An investigation of the use of the alum for defluoridation of drinking water in developing countries, Masters thesis report, University of North Carolina, 1984.

Robinson, R. B., Reed, G. D. and Frazier, B., Iron and manganese sequestration facilities using sodium silicate, *J. AWWA*, 84, 77, 1992.

Rubel, F. and Wooseley, R. D., The removal of excess fluoride from drinking water by activated alumina, *J. AWWA*, 71, 45, 1979.

Russell, L., Recent pilot studies on the removal of iron and manganese, *Proceedings of the AWWA California-Nevada Section Conference*, San Jose, 1977.

Seppanen, H. T., Experiences of biological iron and manganese removal in Finland, *J. Institute of Engineers (India)*, 6, 333, 1992.

USAID Sri Lanka Project 383-0088, *Design manual D3—water quality and treatment*, National Water Supply and Drainage Board, Sri Lanka, 1989.

USPHS, *USPHS Drinking Water Standard*, United States Public Health Service, 1984.

Vigneswaran, S., Shanmuganantha, S. and Mamoon, A. A., *Trends in water treatment technologies*, ENSIC Reviews No. 23/24, Environmental Sanitation Information Center, Asian Institute of Technology, Bangkok, December 1987.

Vigneswaran, S. and Joshi, D., Laboratory development of a small scale iron removal plant for rural water supply, *AQUA*, 39, 300, 1990.

Viswanathan, M. N., Iron removal studies at Tomago sandbeds, *J. Institute of Engineers (India)*, 70, 6, 1989.

Wong, J. M., Chlorinaton filtration for iron and manganese removal, *J. AWWA*, 76, 76, 1984.

Wu, Y. C. and Nitya, A., Water defluoridation with activated alumina, *J. Environmental Engineering Division, ASCE*, 105 EE2, 357, 1979.

Zainuddin, B. A., Improvement of filtration operation by chlorination—filtration process for iron removal, M.Eng thesis report EV-85-1, Asian Institute of Technology, Bangkok, 1985.

8: Disinfection

CONTENTS

8.1 INTRODUCTION

A safe water supply is the prime concern of any water authority. Safety against disease is of utmost importance in water supply, especially the waterborne, water-related, or water-based diseases such as cholera, typhoid, and many others. Even though a certain degree of disease-causing pathogen removal is achieved in treatment processes such as flocculation and filtration, there still exist a significant number of pathogens, particularly viruses, that are not removed by these processes. This necessitates subsequent treatment to remove pathogens, which is conventionally carried out by disinfection. Disinfection is generally the final treatment process in the water treatment train, so that the disinfection effect persists until the treated water reaches the consumer.

Disinfection can be carried out through both physical and chemical means, but the latter has been used extensively for more than a century. The most popular technique so far is chlorination, but other techniques such as ozonation, UV disinfection, etc., are also gaining prominence. The objective of disinfection is to destroy pathogens in water and provide some additional protection against subsequent contamination. The following free chlorine residuals are recommended at different locations in the water system (NWS & DB, 1989).

Clear water/chlorine contact tank: 0.8 to 0.9 mg/L
Standposts: 0.2 mg/L
Distribution dead ends: 0.1 mg/L

In general, disinfection of water is influenced by the following factors (IRC, 1981):

1. Nature and number of pathogens to be destroyed
2. Type and concentration of disinfectant
3. Temperature of water
4. Time of contact
5. Physical and chemical properties of water to be treated
6. pH (alkalinity/acidity) of water
7. Mixing

Chlorine dissolves more readily in cold water (double at 15°C vs. 30°C). In the liquid form it has a yellow (amber) color and a specific gravity of 1.4. The gas is toxic and is irritating to eyes, impairs respiration, and can cause death at high dosages. Chlorine gas and chlorine solutions are highly corrosive and should be conveyed in plastic pipelines.

8.2 CHLORINATION

8.2.1 Chlorine Requirement and Feeding Arrangements

Chlorination can be achieved in two ways: direct solution feed, where chlorine gas is fed directly into the treated water, or else using hypochlorite salts such as calcium hypochlorite $[Ca(OCl)_2]$ and sodium hypochlorite [NaOCl]. Hypochlorite tablets are used for emergency disinfection in times of disaster, when normal water treatment is not operating.

Chlorination with gas feeders may be direct-feed pressure injection of gas metered under positive pressure or by solution feed in which the gas is metered through an orifice or rotameter under vacuum by an ejector drawing gas through the feeder into the solution line. These two feed systems are discussed in detail by Schulz and Okun (1984). In this section, only the simple dosing method of "bleaching powder dosing," which is used in small community water supplies, is discussed.

To provide sufficient time for disinfection, a chlorine contact tank (which may be the filter clear water well) must be provided to give at least 20 minutes detention. The tank should be baffled to prevent short-circuiting. Chlorine solution should be added through a diffuser at the inlet, preferably under turbulent conditions. An example of chlorine contact basin design is provided at the end of this chapter.

The required capacity of the chlorinator will vary with the chlorine dosage needed for disinfection and the amount of residual to be maintained in the distribution system. The dosage will vary according to the factors listed in Section 8.1. For design purposes, a chlorine dosage of 2 mg/L can be used for clear water of less than 10 NTU and with iron and manganese concentrations less than 0.3 mg/L. In view of the fact that ammonia removal, phenol destruction, hydrogen sulfide removal, Fe and Mn removal, etc., imparts a chlorine demand on a raw water, a chlorine demand test should be carried out on the particular raw water to be disinfected, as a predesign exercise.

8.2.2 Bleaching Powder and Dosing Equipment

The dosing of disinfectant is performed in various ways, depending on the availability of the disinfectant, economy, capacity, type of raw water source, etc. Mechanical and electrical dosing pumps are used for disinfectant dosing purposes. The use of dosing pumps in developing countries has to be reconsidered, because of the problems encountered with their operation and maintenance.

To facilitate operation with least maintenance when forms of chlorine other than that of gaseous chlorine are to be used, the dosing equipment will have to be changed appropriately. One example is bleaching powder dosing equipment used in India (Unvala,1979). Here bleaching powder $[Ca(OCl)_2]$ is dosed into the water stream without any utilization of external energy.

Calcium hypochlorite is available as tablets or as a white granular powder with an average of 70% available chlorine. In chlorinated lime, a form of calcium hypochlorite, the available chlorine content is 25–37% (Bryant et al., 1992). The bleaching powder is dissolved in water in desired concentration to prepare a stock solution. The solution is allowed to settle for a few minutes to allow precipitation of the calcium carbonate formed due to the chemical action. If the water is very hard, then soda ash must be added to form more stable sodium hypochlorite $(NaOCl)$ so that the calcium hypochlorite will not be wasted in the reaction to remove the hardness. $NaOCl$ is commercially available as a 10% liquid.

Chemical dosing equipment usually consists of the following:

- solution preparation tank
- pressure vessel
- withdrawable injection arrangement
- orifice plate or venturi tube

The solution preparation tank is where the bleaching powder solution is prepared. Only the supernatant of this tank should be utilized to avoid clogging problems in the system. The pressure vessel consists of a welded steel cylinder with a removable flange cover, a drain, an outlet valve, and a flexible bag made of a material such as plastic. The pressure vessel is used to dose the stock solution from the solution preparation tank. A withdrawable injection arrangement, consisting of an isolating valve, injects the disinfectant dosage into the pipeline from the pressure vessel. The

Table 8.1 Typical Design Parameters of
Chlorine Powder Dosing Equipment (Unvala, 1979)

Parameter	Value
Size of doses (capacity)	36–216 L
Working pressure of pressure vessel	1.7 bar
Testing pressure of pressure vessel	3.4 bar
Differential pressure normally employed	≈ 2 m·head water
Rate at which raw water can be disinfected	<750 L/min

orifice plate is used to regulate the water coming into the pressure vessel through its inlet. Water in a known amount is used to regulate the dosage from the pressure vessel.

The dosing mechanism is simple and straightforward. First, the pressure vessel is isolated from the mainstream pipe and the stock solution is allowed to fill into the bag. The water surrounding the bag gradually drains through the drain valve as the stock solution fills the bag. Once the bag is full, the filling valve is turned to the running position so that the interior of the bag is connected to the downstream of the mainstream pipe through the injection pipe and fitting. Then the valve in the upstream connection is opened to let water into the pressure vessel, which increases the pressure outside the bag; this squeezes the bag so that the stock solution is gradually dosed into the mainstream. The injection fitting is a non-return valve type that allows only the disinfectant to flow into the mainstream. The dosage gets adjusted according to the flow in the mainstream and needs to be calibrated in advance. The stock solution must be fed into the bag periodically.

Extensive application of chlorine powder dosing equipment is found in India (Unvala, 1979). Typical values are listed in Table 8.1. Some advantages and disadvantages of chlorine powder dosing equipment are given in Table 8.2. However, it must be remembered that bleaching powder is relatively unstable and has a very short storage life; also, the available chlorine in bleaching powder is only 25–30% by weight.

Table 8.2 Relative Merits of Chlorine Powder Dosing Equipment (Vigneswaran et al., 1987)

Advantages	Disadvantages
No danger of suffocation due to leaks in enclosed areas, e.g., in basements, etc.	Generally not suitable for treatment plants with capacities greater than 45,000 L/h.
No need to have chlorine cylinders which are scarce in developing countries.	Bleaching powder forms precipitate with hardness and other chemical reactants and the precipitate may clog filters, deposit in pipelines, and affect fittings.
Can be used even at extreme temperatures such as less than 10°C and more than 50°C at which chlorine gas becomes unsuitable due to "chlorine ice" formation and due to leakage respectively.	Due to corrosiveness of the bleaching powder, the dosing equipment must be made of non-corrosive materials.
No mechanical or moving parts and therefore, least operation and maintenance problems.	Bleaching powder is relatively unstable and has a short storage life.
Can be manufactured in developing countries with locally available materials.	Bleaching powder has only 25% chlorine content, which is rapidly reduced with time.
Very marginal cost is involved and therefore, quite economical.	

Simple, Constant Head Hypochlorite Feeders

For small water treatment plants, chlorine can be drip-fed into the water supply utilizing a constant head bottle as shown in Figures 8.1 and 8.2.

Floating Bowl Chlorinator to Feed the Chlorine Solution at a Constant Rate

The simple chlorinator shown in Figure 8.1 can be constructed using any of the following:

1. A large-mouthed plastic bottle with the bottom removed
2. A 6" diameter PVC pipe with one end cemented or welded
3. Any available bowl with a hole cut for a rubber stopper

The central 3 mm outer diameter tube with a nylon string passing through it serves to position the bowl in the center of the tank and prevents it from tipping over. Placing stones in the bowl helps to keep it upright. If glass tubes of 3 mm and 6 mm outer diameter are not available, ballpoint pen casings can serve the same purpose. The two tubes on either side of the central guide can be moved by sliding to a desired height above the rubber stopper. The rate of drip feed is controlled by the difference of head between the level of the 3 mm O.D. tube and the level of the liquid in the tank. If necessary, the outlet of this tube can be narrowed down to reduce the rate of flow to the required level.

Plastic Can Chlorinator *for wells (Brunnen)* ?

The drip-feed chlorinator shown in Figure 8.2 can be used for the disinfection of wells. The container is placed on the parapet of the well, with the outlet tube extending right into the well, dipping into the water. Clogging of the tube may take place due to calcium carbonate deposits that are formed when the bleaching powder solution comes into contact with atmospheric carbon dioxide (IRC, 1981). The solution feed rate is controlled by adjusting the orifice diameter of the stem of the tee. The diameter may be decreased by drawing out the glass stem over a gas flame and nipping it off by trial and error until a satisfactory feed rate is achieved. Alternatively, a pinch-cock clamp may be used in the outlet (flexible) plastic line. For a given orifice diameter and depth below float, the unit will discharge at a near constant rate due to the constant hydraulic head maintained above the orifice by the movement of the float as the solution is discharged.

A similar system has been used successfully in Sudan for temporary chlorination of rural water supplies. (McJunkin, 1967). Here, chloride of lime ($CaOCl_2$) was used as the disinfectant and the stock solution was 20 g/L (100 g for the 5-L can). This was mixed in a bucket and kept for a few hours before pouring into the chlorinator. The inactive sediment was excluded and thrown away. Mixing and standing of solution in alternate buckets was timed in accordance with the quantity

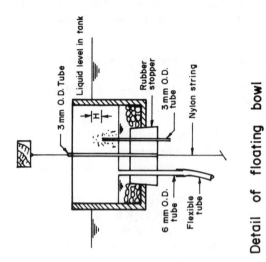

Detail of floating bowl

FIGURE 8.1   Floating bowl chlorinator. (NWS & DB, 1989)

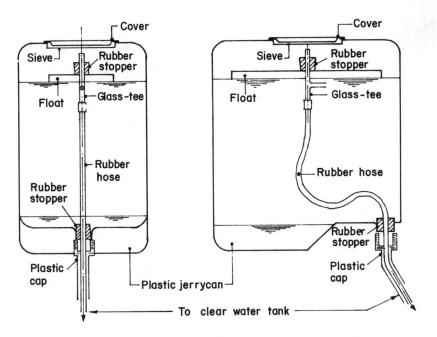

FIGURE 8.2 Plastic can chlorinator. (IRC, 1981)

of water to be treated. Tomorrow's first bucketful is mixed today and stands overnight.

8.2.3 Mixing and Contact Time

Rapid mixing of chlorine at the point of application is important. Chlorine should be added with a large amount of dilution water through a full channel or pipe diameter diffuser submerged to the maximum depth available. Chlorine applied prior to a weir will be driven off by the aeration effect of the hydraulic jump. Chlorine added to the suction side of a pump or the upstream side of a valve can cause severe corrosion problems to brass unless applied at a sufficient distance upstream to obtain full dissolution.

The contact time required is a few minutes for free chlorine to many hours for combined chlorine. A minimum contact time of 20 minutes should be provided either in the pump sump, reservoir, and/or in the pipeline before the water reaches the first consumer. If it is not possible to have a contact time of 30 minutes before the water reaches the first consumer, the consumers concerned should be warned to consume water only after a storage of 30 minutes or more, and they should be ready to accept the use of water with higher doses of chlorine.

8.2.4 Disinfection of Mains and Storage Tanks

Table 8.3 gives the bleaching powder requirements for disinfection of mains and storage tanks. This table is based on a dose of 50 mg/L available chlorine.

Table 8.3 Bleaching Powder Requirements for Disinfection
of Mains and Storage Tanks (NWS and DB, 1989)

Pipe (size in mm)	Bleaching powder requirement
50	0.5 kg/km
65	0.8 kg/km
80	1.2 kg/km
100	1.9 kg/km
150	4.3 kg/km
200	7.4 kg/km
Storage tanks	0.24 kg/m^3

8.2.5 Monitoring

Constant monitoring of effective chlorination by checking the presence of chlorine
at the end and other parts of the distribution system is important, because the quality
of water received at the source changes with weather and season. Regular monitoring
will also provide information on possible sources of contamination. The dosages
should be adjusted to meet the chlorine demands.

8.3 ALTERNATIVE DISINFECTANTS

The question always arises as to why an alternative is necessary to the conventional
techniques used for decades, such as chlorine disinfection. In this case, it is the
undesirable effects caused by chlorine that has prompted researchers to look for
alternatives. Some of these undesirable effects and hazards are as follows:

1. By-products of chlorination have been identified as possessing carcinogenic properties (Bull, 1982).
2. Trihalomethanes (THMs) are considered to be carcinogenic. Trihalomethanes are the products of the reaction between chlorine and natural
   organics such as fulvic acid, humic acid, etc., and synthetic organic matter
   present in the water.
3. The taste produced by chlorination is not aesthetically desirable.
4. Chlorine is unnecessarily wasted through its reaction with ammonia present
   in the water.
5. Careful control of dosage is necessary for safe and effective chlorination.
6. Storage and handling of chlorine are not easy operations, particularly in
   developing countries where liquid chlorine evaporates due to high temperatures (Grombach, 1983).

These undesirable effects can be overcome by choosing a proper alternative method,
such as

1. changing the point for chlorine application such that chlorine is applied
   after substantial removal of THM precursors

2. implementing some process changes to minimize the formation of chlorinated organic and to optimize the precursor removal
3. using an alternative disinfectant
4. removing the THMs after they are formed

Selecting a suitable alternative to chlorine seems to be gaining popularity. Alternatives that have been tried either on laboratory scale or full scale are

- chloramine (NH_2Cl and $NHCl_2$)
- sodium hypochlorite
- chlorine dioxide
- ozone
- potassium permanganate
- ultra-violet rays
- boiling

Out of the above alternatives, chloramine, ozone, and chlorine dioxide have found relatively wider application in replacing chlorine. Even ultraviolet radiation is finding noticeable usage. The above alternatives are discussed in the following sections.

8.3.1 Chloramine

It is the avoidance of THMs that is of great interest in selecting chloramine as a disinfectant, because chloramine impedes the formation of THMs. In a survey conducted in 1984 in the United States, none of the utilities using chloramine reported any adverse effects on bacteriological quality (Staff, 1985a). A recent survey on the occurrence of disinfection by-products indicated that the occurrence of almost all of the by-products was significantly reduced in the systems using chloramine as opposed to chlorine (Bryant et al., 1992). The advantages and disadvantages of using chloramine as a disinfectant are listed in Table 8.4.

Table 8.4 Relative Merits of Chloramine as a Disinfectant (Adapted from Vigneswaran et al., 1987)

Advantages	Disadvantages
Insignificant formation of trihalomethanes (THMs) and other disinfection by-products.	Not as effective as chlorine in deactivating bacteria, viruses, and *Giardia*.
Eliminates certain taste and odor conditions associated with chlorine.	May produce chlorinated phenols, which gives taste to water.
More stable residual in water distribution system.	May produce gas poisoning hazards similar to that of chlorine.
Introduction of chloramine is simple and similar to that of chlorine.	Uncontrolled dosage of ammonia could lead to nitrification problems.
More stable than chlorine (Kreft et al., 1985).	Takes longer time than chlorine for effective disinfection.

Chloramine has been used extensively on a large scale in the United States since 1930 (Staff, 1985a). In fact, the Denver (Colorado) Water Department has been using chloramine as a disinfectant since 1916 (Dice, 1985). It was also found that the distribution system was free of any bacterial slime growth because of the persistence of chloramine residuals in the water. The system uses chlorine and ammonia in a pretested ratio (1:1) to dose into the water. Ammonia is dosed first and thus removes the precursors to a large extent.

The Metropolitan Water District of Southern California changed to chloramine from chlorine in November 1984, also in an attempt to prevent the formation of THMs. The water supply serves about 13 million people and amounts to about 6113 MLD. Ammonia and chlorine are added to the effluent from treatment. The treatment process consists of coagulation and flocculation, sedimentation, filtration, and pH adjustment (Kreft et al., 1985). A number of issues were considered prior to the change to chloramine:

1. chlorine and chloramine chemistry
2. disinfection capability
3. taste and odor
4. growths (e.g., algae) in uncovered reservoirs
5. effectiveness of residuals in the distribution system
6. health effects of chloramine
7. implications to public by changing to chloramine usage
8. technological feasibility
9. economy

On the whole, the changeover was found to be feasible and hence was carried out in 1984.

8.3.2 Chlorine Dioxide

Of all the disinfectants tried so far, chlorine dioxide (ClO_2) has been found to be the most efficient but the most expensive (Grombach, 1983). It must be produced on-site and can be stored for a few hours only (see Figure 8.3).

Chlorine dioxide gas is yellow-green in color and soluble in water upto 3 g/L. Its boiling point is 11°C and the melting point is –59°C. It has an unpleasant odor and is irritating to the respiratory tract in concentrations above 45 ppm in air. Chlorine dioxide is manufactured by mixing a solution of sodium chlorite ($NaClO_2$) with either gaseous chlorine or hydrochloric acid in a solution of low pH (Grombach, 1983). Since chlorine dioxide is quite unstable, care must be taken to avoid possible explosions of ClO_2 gas by preventing the build-up of its concentrations beyond critical levels. Experience shows that use of ClO_2 as a disinfectant needs more care and control than other disinfectants (AWWA Committee Report, 1982). Some advantages and disadvantages of chlorine dioxide as a disinfectant are listed in Table 8.5.

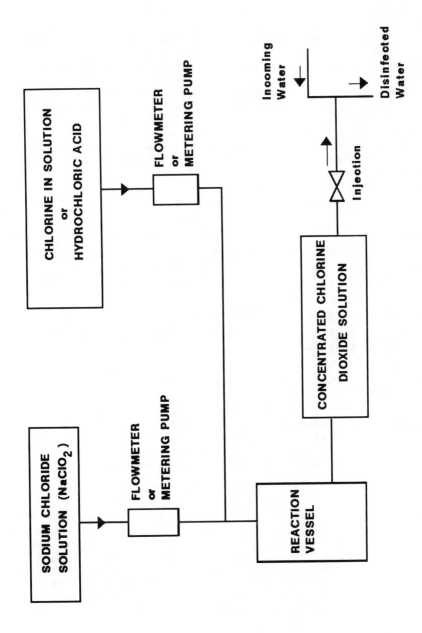

FIGURE 8.3   General diagram of the principle of generating chlorine dioxide.

Table 8.5 Relative Merits of Chlorine Dioxide as a Disinfectant (Adapted from AWWA Committee Report, 1982; Grombach, 1983)

Advantages	Disadvantages
The most effective disinfectant.	Chemical and equipment costs are substantially
Has a very high oxidizing power and thus a low reaction time (only a few seconds).	high compared with chlorine and must be produced on-site.
Forms residuals persistent throughout the distribution system.	The by-products chlorate and chlorite may be toxic.
More effective bactericide and viricide at pH 8.5–9.0, and water is frequently disinfected at this pH rather than pH 7.0.	Contact between chlorine dioxide and organic substances such as wood may lead to self-ignition. The manufacturing equipment are complicated and are subject to corrosion.
Promotes better coagulation and flocculation, especially when colored compounds are present in raw water (Masschelein, 1985).	Application is still limited. Reaction product poses antithyroid activity and by-product ClO_2 may cause hemolytic anemia
Limited tendency for formation of THMs and other by-products that occur with chlorine and chloramine.	(Bull, 1982).

Application of chlorine dioxide as a disinfectant seems to be limited. It is used extensively in Brussels, Belgium (Masschelein, 1985), but its application in the United States seems to be quite limited (AWWA Committee Report, 1982). A summary of Belgian plants using chlorine dioxide as a disinfectant is given in Table 8.6.

The data in Table 8.6 show that water from different sources and of different characteristics can be effectively disinfected using chlorine dioxide, but efficient and safe treatment requires substantial sophistication. In particular, the production of ClO_2 is not a simple process.

Table 8.6 Application of Chlorine Dioxide as Disinfectant in Belgium (Masschelein, 1985)

Plant	Details
Yvoir plant	Treatment of water from River Meuse which is polluted by human sewage, humic substances and the infusion of hard groundwater. First applied in 1966. Capacity of the plant is 18,000 m³/d. The reactor is set to release more than 0.6 g ClO_2/s or 3 g ClO_2/m³ of raw water to be treated.
Tailfer plant	Water is also from River Meuse and serves 500,000 people. Water is highly polluted due to untreated sewage discharge. Dosed before coagulation through baffled mixers at dosages of 0.5–1.0 g ClO_2/m³. Mixing basin provides 60 seconds theoretical residence time. ClO_2 also helps in coagulation and in the elimination of iron and manganese. Residual concentration of chlorite is limited to 0.2 g/m³ by posttreatment using ozone. Ozone oxidizes chlorites to produce residual chlorate. Capacity of the plant is 3.0 m³/s.
Pumping stations of St. Martin	Source is groundwater contaminated by hardness and leachate. Capacity is 3,000 m³/d for St. Martin-Lemmens and 7,200 m³/d for St. Martin-Villeret. ClO_2 is injected at pump strainers at 27 m depth. Normal dosages are 53–89 g ClO_2/h. Intermittent operation for more than 10 years, without any failure. Effective residence time is 220 seconds.
Spontin-Lienne plant	Again the source is groundwater; water is impounded in a quarry (capacity 2,000,000 m³). Dosage is 1.9 g ClO_2/m³. The treatment includes micro-screening and filtration. This has been in operation since 1978.

8.3.3 Ozone

The use of ozone as a disinfectant dates back to 1893, when the first water purification plant using ozone was constructed at Oudshoorn, Holland (Hazen, 1992). In Nice, France, a treatment plant has used ozonation since 1906, while upgrading its capacity from 19 to 76 MLD (Cheremisinoff et al.,1976). It is now the most widely used disinfectant, next to chlorine. Over 2000 plants throughout the world use ozonation for disinfection (Bryant et al., 1992).

Ozone is an unstable form of oxygen having three oxygen atoms in its molecule. The oxidizing agent in ozone is the nascent oxygen released when ozone reverts backs to oxygen as it comes into contact with water. It is the strongest oxidant of the common oxidizing agents. Ozone is manufactured by the passage of air or oxygen through two electrodes with high, alternating potential difference. Ozone is produced at the treatment plant itself, with the use of ozone generators ranging in capacity from 6.6 to 120 g/h. Ozonation systems are generally divided into four components: the feed gas preparation system, the ozone generators, the contractors, and the off-gas destruction systems.

Although ozone was first used as primary disinfectant, it has proven to be more effective in preozonation disinfection (Staff, 1985b). In preozonation, ozone is dosed first to oxidize most of the organics including microorganisms, and at the outlet of the plant minute quantities of chlorine are added, mainly to keep a residual throughout the distribution system. It should be noted that ozone does not provide a residual such as chlorine. The advantages and disadvantages of ozone as a disinfectant are listed in Table 8.7.

Table 8.7 Relative Merits of Ozone as a Disinfectant (Adapted from Grombach, 1983; AWWA Committee Report, 1982; Cheremisinoff, 1976: Vigneswaran et al., 1987)

Advantages	Disadvantages
Has strong oxidizing power and requires short reaction time, which enables the germs, including viruses, to be killed within a few seconds.	Toxic (toxicity is proportional to concentration and exposure time).
Produces no taste and odor.	Cost of ozonation is very high compared with chlorination.
Provides oxygen to the water after disinfecting.	Installations are rather complicated.
No chemicals are necessary.	Ozone-destroying device at the exhaust of ozone-reactor is needed to prevent toxicity and fire hazards.
Reacts readily with all organic matter and removes them.	May produce undesirable aldehydes and ketones by reacting with certain organics.
Most of the ozone decays rapidly in water, avoiding any undesirable residual effects.	No residual effect is present in distribution system, thus requiring postchlorination.
Removes color, taste, and odor.	Much less soluble in water than chlorine and, therefore, special mixing devices are necessary.

The alternative use of ozonation has generated much interest because of its ability to avoid the formation of halogenated organics, inherent in the practice of chlorine treatment. However, raw water quality is seen to significantly affect ozonation results, and could lead to the formation of other undesirable by-products (Bryant et al., 1992).

Brominated by-products are seen to be of major concern in source waters containing bromide (Ferguson et al., 1991). A study on the application of ozonation for the removal of disinfection by-products (DBPs) and natural organics, in addition to disinfection, indicates that in general, ozone does not affect significantly the concentration of precursors of DBPs such as TOCs and reduces the concentration of DBPs mainly by reducing the amount of chlorine needed (Ferguson et al., 1991). Ozonation produces its own by-products, such as aldehydes, ketones, and carboxylic acids. It was found that biological filtration after ozonation, but before secondary disinfection, helps in removing these undesirable by-products (Ferguson et al., 1991).

It is interesting to note that the Los Angeles Aqueduct Filtration Plant, with a capacity of 227 MLD, incorporates preozonation disinfection in its treatment train (Monk et al., 1985). The sources of water are the streams and wells in the Owens Valley and Mono Basin. Chlorine is added at the end of the treatment train for the maintenance of residuals in the distribution system. The treatment train also includes

1. coagulation using ferric chloride as coagulant with cationic polymer as flocculent;
2. tapered flocculation; and
3. a filtration unit using anthracite coal medium.

Preozonation was designed to operate at a maximum ozone dosage of 1.5 mg/L. The decision to use preozonation was based on the following factors:

1. aids the flocculation process and aids higher filtration rates
2. found to be more cost-effective over a 50-year life cycle than chlorine or chlorine dioxide
3. found to reduce flocculent dosages
4. found to produce consistently lower filtered water turbidities than prechlorinated water at the same filtration rates

In addition, the advantages listed in Table 8.7 also contributed toward the selection of ozone for the Los Angeles plant. All the above results were confirmed by pilot-scale studies (Monk et al., 1985).

Ozonation of the public water supply is also carried out in Manroe, Michigan, where it has been operating successfully for more than six years. Due to the locality, the raw water from Lake Erie is of a very bad quality. It has a bad taste and odor. The quality changes substantially from season to season. Hence, after considering the advantages of ozone over chlorine in such a situation, ozonation was selected as the disinfecting method. The water flow is 68 MLD and ozone dosage is a maximum of 3 mg/L. Of all the methods tried, ozonation was the only method found to be successful in removing bad taste and odor completely (Le Page, 1985). Some side benefits realized from ozonation as reported by Le Page are

1. distinct cost saving
2. elimination of the use of powdered activated carbon

3. elimination of the use of potassium permanganate
4. 29% reduction in chlorine application
5. improved coagulation and settling with alum
6. extended filter runs
7. reduction in frequency of filter backwashing and backwash water consumption
8. reduction in wastewater treatment cost
9. partial to complete disinfection
10. stabilized chlorine residuals
11. substantial savings in labor
12. rapid destruction of cyanide
13. deodorized sludge sumps
14. elimination of consumer complaints

8.3.4 UV Radiation

Among the disinfectants under consideration, only ultraviolet rays have some limited application in the field. They are suitable only for plants with a capacity up to 10,000 m^3/d. The complicated and sophisticated equipment and the absence of sufficient experience in treatment plants still makes them unsuitable in developing countries despite their advantages. Also, their effectiveness is hindered by turbidity and other colloids, thus making their use on surface waters impractical (Grombach, 1983). The combination of UV with ozone or hydrogen peroxide has been found to be effective in degrading organics. Guidelines for the design of UV disinfection systems are given in Thampi (1990).

8.3.5 Other Alternatives

The other alternatives for disinfection are still the subject of research and have generally been found to be unsuitable for water treatment on a commercial scale. But, depending on the cost of the disinfectant and the simplicity of the method, they may be tried in developing countries. For example, sodium hypochlorite is relatively safe to handle and requires very simple dosing equipment, usually with no mechanical devices. But it takes a lot of time to act and is consumed in large quantities. The solution also loses its disinfecting power by more than 50% within a few weeks. Destruction of viruses is also not confirmed.

Another method that is gaining ground is the electrochemical treatment of water (Patermarakis and Fountoukidis, 1990), which can be used to produce many different disinfective agents that are extremely effective. The main advantage of this method is that disinfective chemicals are produced on-site, in the treatment systems, removing the need for transport and storage of chemicals. Technological and economical improvement will be required before electrochemical treatment can be applied on a large scale.

EXAMPLE 8.1

Calculate the number of containers required for 4-week chlorine supply in a water supply scheme of 4.0 m³/s flow (maximum daily flow). Maximum chlorine feed rate is 10 mg/L.

Solution

$$\text{Maximum chlorine usage} = 4 \times 10 \times (24 \times 60 \times 60) \times 10^{-3}$$
$$= 3456 \text{ kg/d}$$

$$\text{Maximum chlorine requirement in 4 weeks} = 3456 \times 4 \times 7 = 96{,}768 \text{ kg}$$

If 1-ton chlorine cylinders are used to provide the chlorination (1 ton of chlorine cylinder will consist of 907 kg of chlorine), then the number of containers required in 4 weeks = 96,768/907 = 107 cylinders of 1-ton size.

EXAMPLE 8.2

Calculate the dimensions of a chlorine contact basin that has four-pass-around-the-end baffled arrangement (see Figure 8.4). The contact time at 1 m³/s flow is 20 minutes. The clear width of the basin is 3.0 m, and depth is 4 m.

Solution

$$\text{Volume of a chlorine contact basin} = 1 \times 20 \times 60$$
$$= 1200 \text{ m}^3$$

Therefore, total length of the pass-around-the-end baffles = 1200 m³/(3 × 4 m) = 100 m.

$$\text{Approximately, } 4L = 100$$
$$L = 25 \text{ m}$$

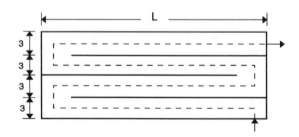

FIGURE 8.4   Chlorination contact basin.

EXAMPLE 8.3

Design a chlorine holding tank to hold the stock solution of disinfectant. Assume a chlorine demand of 10 mg/L including residual chlorine requirement. Commercial NaOCl of 98% purity with appropriate dilution was used as disinfecting agent. Maximum daily flow is 1579 m³/d.

Solution

Chlorine demand per day = 10 mg/L × 1579 m³/d = 15.79 kg/d.

Assuming 35.5 g of Cl is available in 74.5 g of NaOCl, quantity of NaOCl required = 15.79 × (74.5/35.5) = 33.14 kg/d.

Assuming 98% purity, commercial NaOCl demand = 33.14/0.98 = 33.81 kg/d ≈ 34 kg/d.

Assuming stock solution of 5% NaOCl in the chlorine holding tank, amount of stock solution = (34 kg/d)/0.05 = 680 kg/d ≈ 680 L/d.

Assuming that stock solution will be prepared twice a day, volume of the chlorine holding tank = 680/2 = 340 L = 0.8 × 0.8 × 0.55 m³ = 800 mm × 800 mm × 550 mm.

REFERENCES

AWWA Committee Report, Disinfection, *J. AWWA.*, 74(July), 376, 1982.

Bryant, E. A., Fulton, G. P. and Budd, G. C., *Disinfection Alternatives for Safe Drinking Water*, Van Nostrand Reinhold, New York, 1992.

Bull, R. J., Toxicological problems associated with alternative methods of disinfection, *J. AWWA*, 74(7), 642, 1982.

Cheremisinoff, P. N., Vakut, J., Wright, D., Fortier, R. and Maglioro, J., Potable water treatment: Technical and economic analysis, Chapter 8, *Water and Sewage Works*, 123(Oct.), 60, 1976.

Dice, J. C., Denver's seven decades of experiences with chloramination, *J. AWWA*, 77(2), 34, 1985.

Ferguson, D. W., Gramith, J. T. and McGuire, M. J., Applying ozone for organics control and disinfection: A utility perspective, *J. AWWA*, (May), 32, 1991.

Grombach, P., Disinfection in water treatment plants, *Asian National Development*, 32(April), 1983.

IRC—International Reference Center for Water Supply and Sanitation, *Small Community Water Supplies*, Technical Paper Series 18, Rijswijk, The Netherlands, 1981.

Kreft, P., Umphres, M., Hand, J. M., Tate, C., McGuire, M. J. and Trussell, R. R., Converting from chlorine to chloramines, *J. AWWA*, 77(1), 38, 1985.

Le Page, W. L., A treatment plant operator assesses ozonation, *J. AWWA*, 77(8), 44, 1985.

McJunkin, F. E., *Temporary Chlorination of Small Water Supplies*, AID-UNC/IPSED Series Item No. 16, USAID, 1967.

Masschelein, W. J., Experience with chlorine dioxide in Brussels; Part 3: Operation case studies, *J. AWWA*, 77(1), 73, 1985.

Monk, R. D. G., Yoshimura, R. Y., Hoover, M. G. and Lo, S. H., Purchasing ozone equipment, *J. AWWA*, 77(8), 49, 1985.

NWS & DB, *Design Manual D3*, National Water Supply & Drainage Board, Sri Lanka, 1989.

Patermarakis, G. and Fountoukidis, E, Disinfection of water by electrochemical treatment, *J. Water Research*, 24(12), 1491, 1990.

Staff, Chloramines: A practical alternative, *J. AWWA*, 77(Jan), 33, 1985a.

Staff, Ozone comes to America, *J. AWWA*, 77(Aug), 33, 1985b.

Schulz, C. R. and Okun, D. A., *Surface Water Treatment For Communities In Developing Countries*, John Wiley & Sons, New York, 1984.

Thampi, M. V., Basic guidelines for specifying the design of ultraviolet disinfection systems, *J. Pollution Engineering*, 65, May, 1990.

Unvala, S. P., *Bleaching Powder Solution Dosing Equipment*, Candy Filter (India) Ltd., 1979.

Vigneswaran, S., Shanmuganantha, S. and Al Mamoon, A., Trends in water treatment technologies, *Environmental Sanitation Review*, 23/24, December, 1987.

Index

A

Activated alumina, 179–180
Activated bauxite, 180
Activated carbon, 33, 177
Adsorption, 33–34
Adsorption-bridging, 43
Adsorption-destabilization, 49
Adsorptive neutralization, 47
Alabama-type flocculators, 28, 68–71, 76
  application, 70
  design parameters, 69–70
  developing countries, use in, 68
  guideline values for, 70
  pipe material, 69
  process description, 69
  short-circuiting, 70
Alum, 172, 176
Alum coagulation, 178
Aluminum chloride, 178–179
Anthracite coal, determination of suitable grain
  size, 148–149

B

Back-mix-type reactor, 47–48, 53–57
  adsorptive neutralization, 47
  impeller, 47
  impeller design, 56
  mixing chamber design, 55–56
  overflow weir, design of, 57
  power requirement, 56
  propellers, 47
  sweep flocculation, 47
  turbine, 47
Backwashing, 121, 127–131, 138, 150–151
  design criteria, 132
  design principles, 130–131
  design values for, 129
  filter bottom, 130
  headloss, 121, 152
  high-rate water, 129
  interfilter, 128, 130
  low-rate water, 129
  method, 138

  orifices, 130
  rate, 150–151
  surface wash, 128
  system, 127
  underdrain, 131
  washwater, 128
  with air auxiliary, 129
  with effluent from other filter units, 128,
    130–131
Bacterial numbers, 21
Bacteriological quality, 19–20
  enterviruses, 19
  opportunistic (bacterial) pathogens, 19
  pathogenic protozoa, 19
  pathogenic vibrios, 19
  salmonellas, 19
  shigellas, 19
Baffled channel flocculators, 62–64, 73
  advantages and limitations, 64
  design values of, 64
  horizontal baffling, 62–63
  vertical baffling, 62–63
Bilharzia, 2
Biofilm, 21
Biological treatment for removal of iron and
  manganese, 173–175
  advantages, 175
  autotrophic microorganisms, 173
  cost, 175
  dry filters, 173
  oxygen admission levels, 174
  trickling filter, 173

C

Calcium phosphate, 177
Camp's condition, 71
Carbonate removal, 165, 167
  aluminum sulfate (alum), 165
  redox potential, 165
Carcinogens, 12
Chemical quality, standards for, 10
Chloramine, 199–200
Chlorination, 192–198
  chlorine mixing and contact time, 197

209